POCKET GUIDE TO ACCOMPANY

COLLEGE PHYSICS

FOURTH EDITION

by SERWAY & FAUGHN

A Capsule of the Textbook and a Notebook with Problem-Solving Hints

V. GORDON LIND

Utah State University

Saunders Golden Sunburst Series

SAUNDERS COLLEGE PUBLISHING
Harcourt Brace College Publishers

Fort Worth Philadelphia San Diego New York
Orlando Austin San Antonio Toronto
Montreal London Sydney Tokyo

Lind: Pocket Guide to accompany <u>COLLEGE PHYSICS</u>, 4/e by Serway & Faughn

ISBN 0-03-010763-6

5678901234 023 98765432

Preface

As students study physics from a textbook they usually go through it very carefully underlining or highlighting the material most likely needed for an upcoming examination. Some students are more skilled than others in selecting the appropriate material, but basically, all students pick essentially (within 10-20%) the same material. As a teacher of physics students it occurred to me that a small booklet, like this one, could be prepared which would be a skeleton of the text with the most important material included, but with generous margins into which notes also could be added as the student listens to a lecture or reads through the text. Therefore, 80-90% of the note taking is already done and each student can then further personalize the notebook.

This study guide is intended to be a miniaturization of the text and, therefore, designed in order of material to follow the main text. Nevertheless, in reducing the size decisions had to be made as to what should be retained and what is less essential. In making these decisions, it is hoped that the message presented is still coherent, understandable and relevant. It was necessary to write the concepts in my own words and style. Therefore, I accept full responsibility for errors in the manuscript and any lack of clarity.

Every chapter has a brief section on the concepts and one on hints to solving the problems. Students need to solve large numbers of problems in order to master physics, and they need to do it on their own. However, sometimes a little hint is needed to get things going and it is in this spirit of helpfulness that the hint section is included.

Serway and Faughn are masterful writers of physics texts and Saunders College Publishing have utilized their talents and the talents of their Editor, John Vondeling, Developmental Editor, Laura Maier, and their staff to create a most beautiful and useful text which the student can use and cherish for years. By having a small summary book the student will have less weight to carry and any abuse that occurs, as wear and tear, will be on this little book, thus preserving the appearance and usefulness of the text. Also the student will be able to carry the essential formulas and concepts in this abbreviated form wherever he/she goes and thus take advantage of opportunities to study physics during available spare moments that would otherwise be lost.

As an author I appreciate the opportunity provided by Saunders College Publishing to write and provide this little book. It has been a lot of work but also fun to write. I appreciate the support and encouragement from John Vondeling and Laura Maier and Saunders College Publishing. I also appreciate Ray Serway and his help and support. I am greatly indebted to Professor Robert P. Bauman who carefully and critically read the manuscript and offered numerous suggestions for clarification and correction of confusing and misleading statements. Finally, but not least I appreciate the support and encouragement of my wife, Linda, and our children, who patiently understood when my time with them was limited.

V. Gordon Lind, Utah State University, Logan, Utah

Prologue

The Study of Physics

How does
Physics relate to
other Scientific
Disciplines?

What is physics? It is a study of the laws and principles of all phenomena of nature. This includes studies of astronomy, chemistry, geology and even biology (to a large extent) and much more. The principal branches of physics studied in an introductory course, such as this one, are:

- Classical Mechanics (The study of matter in motion and its causes, i.e. forces.)
- Relativity (The study of motion and natural laws near light velocity. It also works at low velocities.)
- Thermodynamics (Statistical behavior of motion)
- Electromagnetism including Optics (Detailed effects of electricity and magnetism resulting from one of nature's fundamental force laws.)
- Quantum Mechanics (A study of motion and behavior of matter on the atomic and nuclear level. This also includes most of modern physics.)

We have learned to trust these theories. Discrepancies between established theories and experiments are unexpected. Physics and science in general have to be based on theories, laws and principles that agree with experimental observations and these do. Experimental results in disagreement with established theory are probably wrong, But if discordant results are repeatable and verified, then the theory, no matter how appealing and popular, must be corrected and brought into agreement with experiment.

The power of mathematics and logical reasoning are needed to develop theories. In some cases they are so powerful as to give insights into nature and predict totally surprising outcomes of experimental observation with great assurance. Examples include the prediction of antimatter before its discovery and very recently the prediction of the existence, and even the mass of the top quark.

As students learn to use physical concepts, theories, laws and principles in the solutions of problems of daily life (and textbook problems) and see the agreement with observation, they gain a sense of excitement and power. As you proceed through the course, you too will marvel at your new abilities. Therein lies the "fun" of physics. It is exciting to understand how and why things work and particularly nature.

Finally, as an aid to success in physics (and anything else), remember that *persistence* and *determination* alone are omnipotent and "practice makes perfect".

The Student Use of the Pocket Guide

This pocket guide is intended to help you learn and master physics as presented in Serway's and Faughn's textbook, <u>College Physics</u>. It is a condensed version of the book with the principles, laws, concepts and equations retained. If you were to go through the text and personally identify the most important ideas by underlining and highlighting, and then put these in a separate book, you would, it is hoped, have essentially the same content as you now find in this booklet. However, it is in no way intended to replace the text. The text is much more complete and the source of your study in learning the concepts. The text also contains problem examples, enrichment material, photographs and illustrations, and sets of questions and problems at the end of each chapter as a resource to you to sharpen your understanding.

There are reasonable margins thus making space for some notes. Also, there are blank pages at the end of about half the chapters. Thus, your pocket guide is your notebook for physics. The pocket guide is small enough to be easily carried everywhere you go. You won't need to carry the big text with you and risk losing it or having it damaged. With a one for one mapping of chapters, sections and subsections between the pocket guide and the text, it should be easy to transfer notes and materials collected from one to the other. By having the pocket guide with you, you can study physics at the bus stop, at the ball game, at the restaurant while waiting for your order, while waiting for a friend, or anywhere else when you have a few moments. It is designed for easy access and for quick review of formulas, concepts, principles and ideas. Mark it up. Personalize it. The more you do the more helpful and valuable it will become.

The "Concept Statements and Questions" section is added to help you gauge your understanding of the subject material in each chapter. Learning and understanding concepts is one of the more difficult aspects of mastering physics. Use this section as it best serves your needs. It may be most useful in studying for a quiz or examination.

Also there is a section on "Hints for Solving the Problems". There are many hints provided. Some are similar to what is found in the text itself and there are also several others. Very few problems have hints directed solely towards their solution. It is realized that this section could be greatly enlarged, but that would make the book too large. Besides, there is a separate study guide with hundreds of worked out problems available for this course and the details you find there will take care of your need for seeing problems worked out in detail. Nevertheless, the hints given here may be just what you need to solve some of the problems and please notice, the hints are directed towards the even numbered problems as well as the odd ones. We add it here as an overall help.

As a guide to learning and mastering physics we suggest the following.

1. Read through the entire pocket guide before the class begins or as early as possible. It is short. This will give you a general background of physics and what you will be studying for the year ahead. This overview will help in preparing you for your learning experience. (Reading the entire textbook would also be good if you have the time.)

2. Read the individual chapters again prior to lecture. If time permits, read the textbook chapter, otherwise, at least read the pocket guide chapter. That way you will be prepared for the lecture and know something about it before you hear the llecture.

3. Attend the lecture with your pocket guide in hand. Listen intently to the lecture, watch the demonstrations and any audio-visual material presented, and listen carefully an explanations. Do not take extensive notes because you will miss much of what is going on if you do. But you can jot a few notes down into the margins of your pocket guide of things that are definitely helpful. Label them L: for lecture. The pocket guide should have most of the information already so what you add is just a supplement.

4. Ask relevant questions of the instructor about unclear concepts and add the answers to your notes.

5. Study the textbook chapter intently and completely and transfer any important items you wish to the pocket guide under the same section heading and label it T: for text. That way you can go back to the text to the exact spot for review if needed.

6. Work the assigned problems and any others you have time for. The more problems you work, the better you will learn and master physics. Get help on the problems you can't work from a fellow student, your recitation teacher, study guide or instructor. Make sure you understand not only how the problems are worked but the details of why the method works and even other ways of doing the same problems..

7. As you prepare for an examination, review your problem solutions and the exercises from the text. Review all the material you will be examined on by going through your pocket guide and the notes you have made in its margins. Study the text as needed, especially the material you have made notes about.

8. As you take the exam do so with confidence knowing that you have prepared well and you will surely do well.

9. From time to time make a quick review of all you learned about physics by reading through the pocket guide and text. The learning acquired in which you invested so much time, will be retained, thereby, and enable you to use it in your future classes and in life's experiences. Otherwise, most of it will soon be forgotten.

Table of Contents

1
Introduction

**Measurement
Standards and
Fundamental Math**

Mathematics is a useful tool and language to help summarize and articulate physical phenomena. In many ways it simplifies the concepts and our ways of viewing nature. Mathematical equations also allow for easy manipulation of the variables.

Only length, time, and mass are needed as a basis for talking about classical mechanics.

1.1 Standards of Length, Mass and Time

The SI (System International) units are used predominantly in science and in the text.

Length: One meter is the distance light travels in a time interval of 1/299 792 458 seconds in a vacuum.

Mass: One kilogram (kg) is the mass of a specific platinum iridium alloy cylinder kept in Sèvers, France.

Time: One second is the time for 9 192 631 770 oscillations of the cesium-133 atom.

See Table 1.4 in the text for metric prefixes representing powers of 10 used in connection with specifying distances, masses, times and other physical quantities.

1.2 The Building Blocks of Matter

Bulk matter can be divided into ever smaller pieces until the atom is reached. At that level it is found that various kinds of atoms exist called elements, and to break matter down further breaks up the atom and we no longer have an element. The atom breaks into the nucleus surrounded by a cloud of electrons. The nucleus is composed of neutrons and protons. The ultimate building blocks of matter on the lowest, most fundamental level are not yet known. They may be tiny objects called quarks, which make up the protons and neutrons, and leptons (an electron is a lepton and there are five others). The quarks and leptons are known and they have no radius or structure. Of course the most fundamental particles may be small still.

1.3 Dimensional Analysis

The symbols of length, mass and time are L, M, T. A physically correct equation must have the same combination of dimensions in each term. If the terms differ in dimesnions the equation, you can be sure, is incorrect.

A check on the dimensions, therefore, is very much worth while and is one good check on the correctness of the equation. The terms in the equation may still be incorrect by a dimensionless multiplicative number, however. Finally, one must be consistent in their use of units for each variable in the equation.

1.4 Significant Figures

Significant figures include the first estimated digit as well as those that are known with certainty.

In multiplying or dividing, the answer has the number of significant figures equal to the least number of significant figures in the factors multiplied or divided.

Scientific notation, 1.5×10^3 or 1.50×10^3 is useful in denoting one more significant figure in the second number and is useful whenever terminating zeros of significance are present.

When adding or subtracting we retain as significant the number of decimal places in the term which has the smallest number of decimal places.

1.5 Conversion of Units

We give only the conversion of a few of the more commonly used units of length.

1 mile = 1609 m = 1.609 km,1 ft = .3048 m = 30.48 cm

1 m = 39.37 in. = 3.281 ft,   1 in. = 0.0254 m = 2.54 cm

1.6 Order of Magnitude Calculations

Without calculators one can estimate the answer to a problem by counting powers of ten and roughly evaluating numerical factors. With electronic calculators, one can compute the answer correctly almost as quickly as estimating it by order of magnitude. Nevertheless, one doesn't always have a calculator and it is very much worthwhile to develop the skill of estimating.

Whenever the answer is obtained one should always consider the question "Is the answer reasonable?" The more one does this the greater his/her skill becomes at sensing when things are not right, when a factor has been left out, or if a number has been mis-entered into the calculator, etc. Thus one gains a *physical sense* about physics.

1.7 Mathematical Notation

Mathematical symbols that are frequently used and their meaning are given by:

$\propto$ denotes proportionality, $>$ denotes greater than, while $<$ denotes less than. The symbol $\equiv$ means *defined as,* $>>$ means much greater than and $<<$ means much less than, and $\approx$ means approximately equal. $|x|$ means the absolute value of x.

We represent the change in a quantity with the Δ symbol. Thus, $\Delta x = x_2 - x_1$ is the change in going from x_1 to x_2. The sum of several different values of x, namely

$+ x_2 + x_3 + x_4 + x_5$ can be represented with the Σ symbol, so that

$$x_1 + x_2 + x_3 + x_4 + x_5 = \sum_{i=1}^{5} x_i$$

where the subscript i on the x represents any one of the numbers in the set.

1.8 Coordinate Systems and Frames of Reference

We specify points in space using a coordinate system with a scale on each axis and an origin

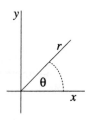

Useful coordinate systems include the cartesian system of mutually perpendicular x-y-z axes and plane polar coordinates r, θ where r is the straight line distance from the origin and θ is the angle at which r is taken .

An x-y position could be specified with the parenthesis around the x and y values, i.e., $(x, y) = (3, 4)$ meaning $x = 3$ and $y = 4$. Similarly for r and θ so that $(r, \theta) = (3, 30)$ means $r = 3$ and $\theta = 30^0$.

The connections between (x, y) cartesian coordinates and plane polar coordinates are

$$x = r \cos \theta, \text{ and } y = r \sin \theta$$

and the inverse, $\tan \theta = x/y$ and $r^2 = x^2 + y^2$

1.9 Trigonometry

Trigonometry is that branch of mathematics that gives relationships between various sides of triangles and also the angles associated. In this course we are mostly interested in right triangles and the relationships known as sine, cosine, and tangent of θ and the Pythagorean formula for the relationship among the three sides. The definitions are:

$$\sin\theta = \frac{\text{side opposite } \theta}{\text{hypotenuse}} = \frac{a}{c}$$

$$\cos\theta = \frac{\text{side adjacent to } \theta}{\text{hypotenuse}} = \frac{b}{c}$$

$$\tan\theta = \frac{\text{side opposite } \theta}{\text{side adjacent to } \theta} = \frac{a}{b}$$

$$c^2 = a^2 + b^2$$

(1.1)

1.10 Problem Solving Strategy

See the six steps in the text for solving problems. They i.e., are useful in developing skills of setting up and solving physics problems which you may not have acquired yet.

The six steps are: (1) read the problem carefully twice, (2) draw a suitable diagram, (3) identify the basic principles involved, (4) select or derive the needed relationship; (5) solve for the desired answer algebraically, and (6) substitute known variables to obtain a numerical answer of the solution.

After obtaining the solution with its units and powers of ten, look at it to see if it is reasonable based on the information given.

1.11 Concept Questions and Statements

1. Mechanics deals with motion and its causes. Why do you suppose that only length, mass, and time are needed for this description since velocities, accelerations and forces are also involved?

2. Why are standards for length, mass, and time needed?

3. It is said that an answer is not meaningful until the units are specified. Why is this so important?

4. The smallest piece of matter for a given element is called *atom*. What is the smallest unit of matter from a compound called? (ans., molecule)

5. Dimensional analysis is useful because the terms in an equation must be all dimensionally the same, and if they are not, then the equation is in error.

6. It is meaningless to give more significant figures in your answer than present in the data given.

1.12 Hints for Solving the Problems

General Hints.

1. Refer to the hints and problem solving strategy given in the text.

2. Read through and thoroughly understand the examples given in the text.

Hints for Solving Selected Problems

1-4 Check the dimensions of each term in the equation. If the dimensions do not agree, answer the questions regarding it. You may have to isolate the variable of interest according to the rules of algebra.

5. In writing your formula do not worry about multiplicative factors. Just make sure the dimensions for h are correct in terms of r and R.

6. g has the dimensions of L/T^2.

7-8. The significant figures you keep need to be rounded off to the nearest last significant digit.

10. Check the significant figure rule for addition.

11-12. Keep only the proper number of significant figures. This will require throwing some calculated figures away.

7-27. In converting units you may substitute equivalents. For example, 1 ft is equivalent to 30.48 cm, so 30.48 cm can be substituted everywhere "ft" appears. Remember to follow the significant figure rules in your final answer.

28-31. Make your best guess to an order of 10 for the variables you put into the calculation. Then carry out the calculation.

32-35. Use the conversion equations between cartesian coordinates and polar coordinates.

36-40. Use the trigonometry relationships provided.

2
Motion in One Dimension

A Simple Study in Kinematics

Mechanics is the study of the motion of objects. To study how an object moves is kinematics and the causes for motion and the changes in motion involves forces and that is known as dynamics. The simplest motion is motion in one dimension. We will consider the motion of point objects called particles in this chapter.

2.1 Displacement

The displacement of a particle is the difference between two positions it occupies. As a particle moves it continually changes position. The displacement Δx is, therefore, $x_f - x_i$ where x_f is the final position and x_i is the initial position.

2.2 Average Velocity

The displacement takes place during a time interval, Δt = t_f - t_i. The ratio of $\Delta x/\Delta t$ is defined as the average velocity (in one dimension and hence we don't use the vector nature of velocity) of the particle. The equation for average velocity, therefore, is

$$\bar{v} \equiv \frac{\Delta x}{\Delta t} = \frac{x_f - x_i}{t_f - t_i} \tag{2.1}$$

And the average velocity is directed along one direction only in this example. The sign of Δt is considered positive since t_f is larger than t_i, so the sign of the average velocity is given by the sign of Δx.

2.3 Instantaneous Velocity

The instantaneous velocity is the velocity at a point and is defined as the limit of the average velocity as Δt approaches zero. So for instantaneous velocity we get:

$$v \equiv \lim_{\Delta t \to 0} \frac{\Delta x}{\Delta t} \tag{2.2}$$

In a plot of x versus t, the slope of the curve at any point represents the instantaneous velocity at that point. The sign of the slope also gives the sign of the velocity. The magnitude of v is known as the instantaneous speed of the particle and is always positive.

2.4 Acceleration

A change in velocity divided by the change in time is known as the acceleration. The average acceleration is simply a ratio of Δv to Δt. The limit of the average acceleration as t goes to zero is the instantaneous acceleration. Thus, the average acceleration is

$$\bar{a} \equiv \frac{\Delta v}{\Delta t}$$

(2.3)

INSTANTANEOUS ACCELERATION

The instantaneous acceleration is obtained from

$$a \equiv \lim_{\Delta t \to 0} \frac{\Delta v}{\Delta t} = \frac{v_f - v_i}{t_f - t_i}$$

(2.4)

Increasing velocity implies positive acceleration and decreasing velocity means the acceleration is negative (sometimes called deceleration).

2.5 Motion Diagrams

A plot of x versus t tells us everything about the position of the particle, its velocity at every point and its acceleration at every point. The velocity is given by the slope of the curve and the acceleration is given by the change in the slope.

2.6 One-Dimensional Motion with Constant Acceleration

If the acceleration is constant, i.e., it doesn't change, then the instantaneous acceleration is equal to the average acceleration and vice versa. If we begin our time at $t = 0$, then we can write the instantaneous velocity simply as:

$$v = v_o + at \qquad (2.5)$$

Also, since a is constant, the average velocity is one half the sum of v and v_0. Combining this with our other equation for average velocity gives us

$$x - x_o = \left(\frac{v + v_o}{2}\right)t \qquad (2.6)$$

By choosing the origin at the starting point x_0, then we make $x_0 = 0$ so our equation becomes $x = (v + v_0)t/2$. Then by substituting for v from Equation 2.5 we get the very useful equation of $x(t)$ as it relates to the initial velocity and acceleration.

$$x = v_o t + \frac{at^2}{2} \qquad (2.7)$$

By substituting in for the time in Equation 2.7 from Equation 2.6, the time is eliminated and we get:

$$v^2 = v_o^2 + 2ax \qquad (2.8)$$

2.7 Freely Falling Bodies

The free-fall acceleration near the earth is nearly constant, so freely falling objects are good examples of problems involving constant acceleration in one direction. The motion is in the vertical direction (y axis) and the acceleration is directed downward in the direction of -y. The free-fall acceleration is denoted g and has the value of 9.80 m/s^2.

All of the equations we developed for constant acceleration apply here, but we replace x by y and a by $-g$. For example, Equation 2.8 becomes $v^2 = v_0^2 - gy$.

2.8 Concept Statements and Questions

1. If the velocity is constant, what is the acceleration? Also what does a curve of x versus t look like for constant velocity? (Thus, if acceleration is present, the curve will deviate from a straight line.)

2. In the question above, which way does the curve bend if the acceleration is positive? Which way does it bend if the acceleration is negative?

3. For constant acceleration, what does a curve of v versus t look like? (Ans: straight line) Now it is clear why the average v is simply $(v + v_0)/2$, i.e., it is in the middle. Looking at Equation 8 we see that v has a linear relationship with respect to t. Whenever one variable is linearly related to another variable, the average of the variable can be simply calculated as done here.

4. If the free-fall acceleration is directed downward will the velocity also always be directed downward? Do the velocity and acceleration have to be in the same direction? Explain.

2.9 Hints for Solving Problems

General Hints

1. Be sure to take the difference between two particle positions to get its displacement. Similarly, be sure to take the difference for the elapsed time. Of course, if the initial position is zero or if the initial time is zero, one only needs to take the final values.

2. The formulas for computing average velocity, average acceleration, instantaneous velocity and instantaneous accelerations are valid and true for all situations. However, the equations developed for motion under constant acceleration should only be used when the acceleration is in fact constant. Do not forget this fact.

3. Study carefully the problem solving strategies provided in the text. Also study the example problems in the text.

Hints for Solving Selected Problems

1-5. Remember the definitions of average and instantaneous velocities and apply these definitions in working these problems. Rearranging the equations with the use of simple algebra is sometimes needed.

6. Solve for the time for each of the three segments of the trip, then sum to get the total time.

7-8. Set up the kinematic equations for each moving object and solve the equations simultaneously.

9-15. Remember the statements about velocity on a position vs. time graph and easily answer the questions.

16-22. Use the definition of acceleration and the rules for acceleration on position vs. time graphs.

24-35. Use the one-dimensional kinematic equations for constant acceleration. Chose the equations that deal with the variables of concern in the problem and that contain the variable of interest that you're solving for.

36-45. These problems involve vertical motion with constant acceleration directed downward. Remember that velocity and acceleration do not have to be in the same direction. If the velocity changes from + to -, the acceleration can still be constant.

3
Vectors and Two-Dimensional Motion

Kinematics of a Particle Moving in a Plane

Motion in two dimensions is easily comprehended if one thinks of the motion as being made up of two independent motions, i.e., motion in each dimension will be considered as independent of the other and then combined to make up the total motion. To represent the total motion we will introduce vectors and discuss procedures for adding and subtracting them.

3.1 Vectors and Scalars

A scalar is a quantity completely specified by only a number with appropriate units. A vector is a quantity that has a magnitude and a direction. Vectors are encountered whenever more than one dimension is needed to consider the quantity of interest. We denote vector quantities with bold face type in this book.

3.2 Some Properties of Vectors

Equality of two vectors A and B: A = B means both the magnitude of **A** and **B** are the same and they are in parallel directions. Their locations are not important in making them equal. They could be in entirely different locations.

ADDING VECTORS: C = A + B. Vectors can be added graphically by placing the tail (beginning point) of one vector, **B**, to the head (end point) of the other, **A**. A line connecting the tail of the first, **A**, to the head of the second, **B**, is the resultant, **C**, of the addition which is also a vector. Vector addition may also be accomplished by adding together components as scalars of each vector **A** and **B**, namely $C_x = A_x + B_x$ and $C_y = A_y + B_y$

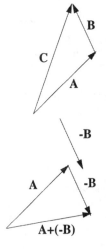

Negative of a Vector: (-**B**), is in the opposite direction to **B** but with the same magnitude.

Subtraction of Vectors: Subtraction is achieved by adding a vector **A** to a negative vector **B**, i.e., **C** = **A** + (- **B**). Displacement is a vector indicating the changes in position of a particle and is, therefore, the difference between the final and initial position vectors, i.e., $\Delta \mathbf{r} = \mathbf{r}_f - \mathbf{r}_i$.

MULTIPLICATION AND DIVISION OF VECTORS BY SCALARS:

Multiplication of a vector by a scalar gives a new vector of different length. It changes the magnitude and hence the length of the vector, but not its direction except if the scalar is negative, in which case the vector points in the opposite direction.

3.3 Components of a Vector

A vector can be represented as the sum of its two components directed along the x and y axes. To specify these components we give their magnitudes as

$$A_x = A\cos\theta$$
$$A_y = A\sin\theta$$

(3.1)

where θ is the angle between a vector directed along x and **A**. Vector **A** is now resolved into its components A_x and A_y. The magnitude of **A** is denoted $|A|$ or simply A (not bold). It is related to A_x and A_y through

$$A = \sqrt{A_x^2 + A_y^2} \quad \text{and}$$
$$\tan\theta = \frac{A_x}{A_y} \quad \text{so} \quad \theta = \tan^{-1}\frac{A_x}{A_y}$$

(3.2)

ADDING VECTORS

Two or more vectors may be added algebraically by first decomposing the vectors into their components, then adding the components algebraically, so $C_x = A_x + B_x$ and $C_y = A_y + B_y$. Of course C and its direction can be obtained from equations like Equation 3.2.

(See the text for problem solving strategy when vectors are involved.)

3.4 Velocity and Acceleration in Two Dimensions

The equation of motion is an expression which gives the position of a particle as a function of time. A trajectory gives an expression for the path the particle takes in the two dimensional plane.

Dealing in two or more dimensions makes the displacement, the velocity and the acceleration variables as vectors since they may have components in each dimension.

The displacement vector is the vector difference between the final position, $\mathbf{r}_f$ and the initial position, $\mathbf{r}_i$, i.e.,

$$\Delta\mathbf{r} \equiv \mathbf{r}_f - \mathbf{r}_i \qquad (3.3)$$

The **average** velocity is the displacement vector divided by the elapsed time,

$$\bar{\mathbf{v}} \equiv \frac{\Delta\mathbf{r}}{\Delta t} \qquad (3.4)$$

and the **instantaneous** velocity can be computed similarly to the average velocity but the elapsed time is allowed to approach zero which also results in the displacement vector becoming very small.

$$\bar{\mathbf{v}} \equiv \lim_{\Delta t \to 0} \frac{\Delta\mathbf{r}}{\Delta t} \qquad (3.5)$$

The **average** acceleration is the change in velocity divided by the elapsed time Δt.

$$\overline{\mathbf{a}} \equiv \frac{\Delta \mathbf{v}}{\Delta t} \tag{3.6}$$

The limit of this ratio as the elapsed time goes to zero is called the **instantaneous** acceleration, namely

$$\mathbf{a} \equiv \lim_{\Delta t \to 0} \frac{\Delta \mathbf{v}}{\Delta t} \tag{3.7}$$

Note, whenever the velocity vector changes either in magnitude or direction, we have acceleration.

3.5 Projectile Motion

A projectile, such as a baseball after it has been hit for a home-run, has vertical motion and horizontal motion. Its path describes a trajectory. It is a good example of two dimensional motion. The horizontal component of velocity is constant so

$$x = v_{x0}\, t \tag{3.8}$$

whereas the vertical component of velocity is affected by gravity which accelerates the object in a downward direction. Thus, $a_y = -g$, and

$$v_y = v_{y0} - gt, \tag{3.9}$$

$$y = v_{y0}t - gt^2/2. \tag{3.10}$$

The initial velocity components can be obtained from the initial velocity and initial angle at which the particle began its trajectory, namely

$$v_{x0} = v_0 \cos \theta_0, \quad \text{and} \quad v_{y0} = v_0 \sin \theta_0. \tag{3.11}$$

Also we have

$$v_y^2 = v_{y0}^2 - 2gy \qquad (3.12)$$

The speed or magnitude of the velocity at any position in the trajectory is given by

$$v = \sqrt{v_x^2 + v_y^2} = \sqrt{v_{x0}^2 + v_y^2} \qquad (3.13)$$

3.6 Relative Velocity

An object in motion observed by someone in motion may appear quite different than it does by an observer at rest or one traveling at a different velocity. A simple way of relating the various velocities is the subscript rule. The rule goes as follows. v_{ab} means the velocity of a with respect to b. So in terms of other velocities we can write

$$\mathbf{v}_{ab} = \mathbf{v}_{ac} + \mathbf{v}_{cb} \qquad (3.14)$$

Note, each velocity is a vector and the subscript starts with the letter that ended the preceding velocity subscript. Also, if the order of the subscripts is reversed there is a change in sign. So, $\mathbf{v}_{cd} = -\mathbf{v}_{dc}$, etc.

3.7 Concept Statements and Questions

1. How do the concepts of velocity and acceleration differ in this chapter from the ideas about these quantities developed in chapter 2?

2. Everything explained about vectors in this chapter could be easily extended to three dimensions by simply adding another component, e.g. A_z.

3. You will notice that the kinematic equations involving two dimensions are completely similar to the equations in one dimension. The total motion is obtained by simply putting the two individual motions together.

4. In a projectile problem the range is usually defined and figured when the vertical height, y, is equal zero. There are always two x values for every y since what goes up must come down and the x position continually changes due to the velocity in the x direction, i.e., v_x. What x corresponds to the other $y = 0$ value?

3.8 Hints for Solving the Problems

General Hints

1. For projectile motion, remember that even though the projectile is accelerating in the vertical direction, this does not affect the speed in the horizontal direction in any way. The horizontal speed remains constant throughout the flight. Likewise, the constant horizontal speed has no effect on what is happening in the vertical direction.

2. The author's way for solving for time in Example 3.5 is perfectly OK, but there may be other ways to think about the problem and some may give an even simpler solution. For example in finding y_{max}, it is simply, $y_{max} = gt^2/2$, where t is the time for half the trajectory, 0.384 s.

(Why is this so?) Of course $v_{y0} = v_0 \sin\theta$ is the initial vertical velocity and the final velocity is the same, but in opposite direction. But these are not needed to solve for y_{max}. The lesson is this: Look for ever simpler ways to solve the problems and try to understand the connection between the various methods and why they each work.

3. Example 3.3 is very good to study to understand the independence of motions in the x-direction and y-direction.

4. The addition and subtraction rules for relative velocity are very useful but must be applied with precision, so study the examples very carefully.

Hints for Solving Selected Problems

1-8. In working these vector problems it is important to make a sketch of all of the displacements given and to identify the unknown quantity that you are solving for on the graph before you attempt the solution. This procedure prepares you for all other physics problems too.

8-16. Draw a sketch of the given displacements and on the sketch identify the unknown you are asked to find. Resolve the given vectors into components and combine the components by adding or subtracting as needed. Solve for the unknowns asked for. In Problem 16, to solve for the unknown displacement you can subtract the known displacements from the resultant.

17. Use the projectile equations for x and for y and eliminate the t. Thus you will have an equation involving x and y in terms of the angle of projection.

18-20. Remember the motion in the horizontal direction is to be treated independently from the vertical motion.

21. The vertical velocities must be equal.

22-24. Consider the vertical and horizontal velocities to be independent, but the time has a common starting point so finding time from one equation and substituting into the other gives the answer.

25-35. Most of these problems can be solved using the relative velocity formula. If the velocities are not colinear, the equation is one of vector addition.

4
The Laws of Motion

**Introduction to a
Study of
Dynamics**

What causes motion to change? i.e., what brings about the change in velocity of an object? Newton helped us to realize that it is always an unbalanced or net force that brings about acceleration and thus it is forces that affect motion. In this chapter we discuss forces and how they bring about changes in motion.

4.1 Introduction to Classical Mechanics

Classical mechanics extends our study of motion from kinematics to include dynamics. In a study of dynamics we learn the connection between forces and accelerations.

4.2 The Concept of Force

What is a force? Newton gave us a way to recognize forces. An unbalanced force (one that is not totally countered by other forces, or the net force) acting on an object causes an acceleration, i.e., a change in its motion. If an object is at rest and remains at rest, or if it is moving with constant velocity and remains that way, then there are no unbalanced forces acting on it. Conversely, if the object's motion undergoes a change, we can be sure there is an unbalanced force acting on it. These two statements constitute the fundamental elements of Newton's first two laws of motion.

Studying the details of the effects of a force enables us to learn all about it.

KINDS OF FORCES

In the macroscopic world we encounter two classes of force, namely *contact* forces and *field* forces. The object producing the force actually touches the object being acted upon in the case of contact forces. In the case of field forces, the force is experienced over a distance such as observed in the case of the force of gravity of the Earth acting on the Moon or a satellite in orbit.

On the atomic level scientists recognize that there are four fundamental forces, namely (1) gravitational (2) electromagnetic, (3) weak nuclear and (4) strong nuclear and they are all field forces. The contact forces experienced in daily life only appear to be contact forces. Actually, most contact forces that we encounter are the electromagnetic force acting as a field force at very small distances so the objects seem to be in contact such as the collision of two cars.

The stretching of springs is used conveniently to measure forces. The length of the stretch is proportional to the force producing the stretch providing the spring's elastic limit is not exceeded. Springs and other devices that follow this are said to be following *"Hooke's Law"*.

Forces are vectors and must be summed as vectors. The vector sum of all forces acting on a body gives the resultant force which is also the unbalanced or net force that causes accelerations.

4.3 Newton's First Law

Newton's first law of motion states: **An object at rest will remain at rest, and an object in motion will continue in motion with a constant velocity (that is, with constant speed in a straight line), unless it experiences a net external force.** The law is sometimes known as *the law of inertia*.

Inertial Reference Frames: An inertial reference frame is one in which *the law of inertia holds,* i.e., the first law of Newton applies. Rotating and other kinds of accelerated reference frames are examples of reference frames that are not inertial. Slowly rotating reference frames, such as the rotating Earth, are approximate inertial reference frames.

MASS AND INERTIA

Mass is used to measure inertia. The greater the mass of an object, the less it will accelerate under the action of an applied force. Remember mass is not the same as weight. Weight is the gravitational force acting on a mass m by some other mass M. Mass is a scalar quantity and is an inherent property of a body, independent of the body's surroundings and of the method used to measure it.

4.4 Newton's Second Law

Newton's second law is stated as: **The acceleration of an object is directly proportional to the resultant force acting on it and inversely proportional to its mass.**

Thus mass, force, and acceleration are all related by the simple formula

$$\sum \mathbf{F} = m\mathbf{a}$$

(4.1)

UNITS OF FORCE AND MASS

The SI unit of force is called the Newton. 1 N = 1 kg m/s^2. If mass is measured in grams and length in centimeters, then the unit of force is dyne = 1 g cm/s^2 = 10^5 N. In the British engineering system the pound is used for force. 1 lb = 4.448 N.

Unbalanced Force **F**

$m\mathbf{a}$

Object accelerates in direction of net force

WEIGHT

Weight is the gravitational force, F_G, exerted on one body of mass, m, by another body of mass, M. On the surface of the Earth we talk about the weight of an object and that is the gravitational force which the entire earth exerts on the object. If released, the object accelerates with an acceleration g because of this force towards the center of the earth. Thus the weight is $w =$

$F_G = mg$. Since the force is proportional to the mass, a spring following Hooke's law will stretch a distance proportional to the weight or the mass and thus springs can be used to measure either weight or mass.

4.5 Newton's Third Law

Newton's third law states: **If two bodies interact, the force exerted on body 1 by body 2 is equal to and opposite in direction to the force exerted on body 2 by body 1.**

The equation representing this statement is:

$$\mathbf{F}_{12} = -\mathbf{F}_{21} \qquad \text{(4. 2)}$$

The action force acts upon the body whose motion we are studying and the reaction force is on the other body producing the action. **One must never assume they are both acting on the same body.**

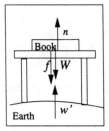

A body of mass m, such as a book, resting on a table top is an interesting example of reaction pairs. The gravitational attraction of the Earth on the book gives a weight force, W, that acts on the book. It doesn't accelerate because the table top exerts an equal and opposite force, n, upward on the book so that the net force is zero. These two forces are not **action-reaction** pairs, however. The reaction to the weight is an equal attraction of the book acting on the Earth (the Earth produced the action force $w = mg$). The reaction force to the force of the table on the object is the equal and opposite force, f, of the book pushing down on the table. Thus, there are *two* action-reaction pairs in this example but only two action forces on the object of interest.

4.6 Some Applications of Newton's Laws

The application of Newton's laws to understand the motion of an object requires that we consider the external forces, since the internal forces cancel in pairs. It is important to identity only those external forces that act on the object and to include all of them. The reaction forces of the object back on the object doing the acting must be excluded.

OBJECTS IN EQUILIBRIUM AND NEWTON'S FIRST LAW

Objects are in equilibrium when Newton's first law applies. The net force on an object in equilibrium is zero. Therefore, the vector sum of all the forces is zero and also the sum of the components in each direction is zero. A recommended strategy in working such problems is to work with each component separately.

When working with tensions in a string, rope, or wire, the tension is a force acting at every point in the wire producing action-reaction pairs. Thus one body may affect another through tension. Nevertheless, the reaction force acts back on the body through the tension.

Study the problem-solving strategy outlined in the text and the examples provided.

ACCELERATING OBJECTS AND NEWTON'S SECOND LAW

In this case the net force acting on an object is not zero. Again, it may be simplest to look at each direction and sum the components of forces in that direction. Careful thought in setting up the problem may enable you to reduce the problem to only one component and thus end up with a scalar equation.

Free body diagrams are very useful in solving dynamic problems. Each object being considered in the diagram should be drawn and all of the forces represented by arrows indicating direction and magnitude (by length). Most problems become much easier to understand and work when this is done.

Again review the problem-solving strategy in the text and the excellent examples. Examples 4.3, 4.4 and 4.5 are quite different from one another and illustrate how to approach and solve many of the problems at the chapter's end and as will appear on tests.

4.7 Force of Friction

Forces of friction are resistive in nature. Thus they are usually opposite to the motion but are sometimes in the direction of the change in motion such as friction holding an object in a circular path. They always act to oppose the relative motion of adjacent surfaces, and may, therefore, cause either a deceleration, as in stopping a car, or an acceleration when drive wheels push against the pavement.

On the atomic level they are electromagnetic in nature and act between the atoms and molecules on the surface of the objects in contact. From a macroscopic view, however, friction is often due to the roughness and irregularities of the surface. Extremely smooth surfaces can have very high attraction for each other and, therefore, large frictional forces between them.

The force of static friction, $\mathbf{f}_s$, resists the applied force and equals it numerically but is oppositely directed, i.e., $\mathbf{f}_s = -\mathbf{F}_a$, up until the object begins to move. Its magnitude is given by

$$f_s \leq \mu_s n \tag{4.3}$$

where n is the force normal to the surface and μ_s is the coefficient of static friction.

The force of kinetic friction, which involves the kinetic friction coefficient, μ_k, is opposite the direction of motion and its magnitude is given by

$$f_k = \mu_k n \tag{4.4}$$

FRICTION AND THE MOTION OF A CAR

A car is propelled forward through friction, namely the friction between the tires and the road. The reaction force, f, to the backward friction action force between the tires and road enable the car to move in the opposite direction. Yes, the friction forces on the tires are opposite to the direction of motion of the tires, but are in the direction of motion of the car and propel the car forward. The car also experiences air resistance which is proportional to its velocity and opposite to the direction of motion. This resistance, $\mathbf{R}$, is given as $\mathbf{R} = -b\mathbf{v}$.

4.8 Concept Statements and Questions

1. Some people misinterpret Newton's first law as a special case of the second law. How do they differ? Why must the first law be established before the second law can be understood?

2. Why are mass and weight considered to be different? In what ways are they the same? It is very important that you understand the distinction.

3. When an object accelerates, only the forces acting on the object are affecting its motion. If that is the case, what do the reaction forces affect? Do reaction forces ever need to be taken into consideration?

4. Under what circumstances would you notice a failure in Newton's first law of motion.

5. Why is an inequality sign used for the equation for static frictional force? When does the equal sign apply?

6. A car turns around a sharp corner. If a frictional force holds it in a circle, is that force static or kinetic? What determines whether the force is static or kinetic?

4.9 Hints for Solving the Problems

General Hints

1. Most of the problems in this chapter are worked by either finding the force once an acceleration is known, or to find the acceleration once the force is known. For

all such problems, Newton's second law is used. If the problem asks for average force, it can be obtained from average acceleration or vice versa. Average acceleration was discussed in the previous two chapters.

2. Some of the problems involve directions of the accelerations and forces not colinear with the usual set of coordinate axes. If you work these problems using components along the axes then the components can be combined to give the final answer. Look for ways of simplifying the problem by reorienting the coordinate axes so that fewer components need to be evaluated.

3. In working with several forces each in a different direction, working with components is usually the easiest way to solve the problems.

Hints for Solving Selected Problems.

1-14. Use $F = ma$ to get F. a can be obtained from kinematics as you learned in chapters 2 and 3. Once you have F, you can determine a for other masses if needed.

15-20. The key here is "equilibrium". The sum of all the forces (including tension) on each object that is at rest or moving with constant velocity, must add up to zero. This enables the unknown forces to be solved from the known ones

21-37. These are mostly $F = ma$ problems so the comments for problems 1-14 apply. You should also apply the problem solving strategies outlined in the text including making a sketch of each object experiencing forces and or motions. From these sketches you will be able to see how the problem is solved.

33-37. Remember that tension acts in both directions along a string, wire, or rope.

39. *M* experiences a weight force so it would be different on the Moon, but *m* experiences a frictional force which also depends on the acceleration due to gravity.

40. From the data given you can determine the acceleration and, therefore, the force and hence the coefficient of friction.

41-59. Use $\mathbf{F} = m\mathbf{a}$ including friction forces and work with x and y components. Remember that if changes in velocity are given over a distance, one can deduce the acceleration from the kinematic equations. Working with components can be helpful. Use sketches and free-body diagrams as needed.

5
Work and Energy

**Making Good Use
of Energy**

Energy comes in a great variety of forms. Its great value to us is its capability to do work, which is one type of energy transfer to a system. Another great quality of energy is its conservation, i.e., the amount of energy in a closed system is always constant even though it can be changed from one form to another. Often what is called work is just changing energy from one form to another. Thus in lifting a brick from one elevation to another we do work on the brick, but the work we do just simply goes into increased energy of position (potential energy) and can be recovered by letting the brick fall.

In this chapter we will restrict our discussions to mechanical energy. We will develop the concept of work and see its relationship to energy and see how that working with energy (a scalar) can make otherwise difficult problems involving forces (vectors) very simple.

5.1 Work

In physics work is defined as a force (or component of force) applied in the direction of motion over a distance, so that

$$W = (F\cos\theta)\,s \qquad \text{(5.1)}$$

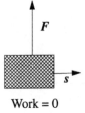

Work = 0

Here s is the displacement. It is clear from this definition that a force applied perpendicular to the direction of displacement does no work since the angle is 90^0 and the cosine of that angle is zero. Thus lifting a heavy object (vertical force, vertical displacement) does work, but then carrying it horizontally across a level room, requires no work. The person carrying the object may exert considerable effort and become exhausted from his/her efforts and thus we may think of work being done from a physiological point of view, but not from physics. We could have put the object on a frictionless cart and moved it across the room with no physiological effort. No work is done because no energy was transferred to the object.

Work is a scalar and its SI unit is newton-meter = N·m. We also call a newton-meter a joule (symbol J).

5.2 Kinetic Energy and the Work-Energy Theorem

Suppose a net constant force is applied to a particle in the horizontal direction. The particle accelerates from speed v_i to speed v_f and undergoes a displacement s. The work done by $\vec{F}_{net}$ is

$$W_{net} = (F_{net})s = (ma)s \tag{5.2}$$

For a constant force we know from kinematics that the distance and acceleration are given by $s = (v_0 + v_f)t/2$, $a_x = (v_f - v_i)/t$, respectively, therefore the work is

$$W_{net} = m\left(\frac{v_f - v_0}{t}\right)\left(\frac{v_0 + v_f}{2}\right)t = \frac{1}{2}mv_f^2 - \frac{1}{2}mv_0^2 \tag{5.3}$$

Thus the net work is simply the increase in kinetic energy which we define as $KE = mv^2/2$.

This important result is known as the work-energy theorem and is equally true even if the force is not constant. It is often convenient to write Equation 5.3 as

$$W_{net} = KE_f - KE_i \tag{5.4}$$

We conclude: **The net work done on a particle by a net force acting on it is equal to the change in the kinetic energy of the object.** (The theorem is valid only if the objects can be considered to be particles.)

5.3 Potential Energy

An object with energy has the ability to do work. The work-energy theorem shows how kinetic energy can change and result in an amount of work being done. Similarly, objects with potential energy can do work by giving up some of their potential energy, which means they change their position. Gravitational potential energy is given by

$$PE = mgy = \text{weight} \times \text{vertical position} \tag{5.5}$$

The weight is fixed but the vertical position is defined only in terms of some origin in a reference frame and that is arbitrary. The potential energy can only be specified when that origin is identified. However, the difference between two vertical positions does have a unique value, as also the difference in potential energy. So $\Delta PE = PE_f - PE_i$, is unique and has physical meaning.

Because the force of gravity is always directed downward, the work done by the gravitational force in moving an object from one position to another is

$$W_g = -\Delta PE = PE_i - PE_f \qquad \text{(5.6)}$$

Thus we observe that the work done by the gravitational force is opposite in sign to the work done against the gravitational field as we would expect.

5.4 Conservative and Nonconservative Forces

Forces can be divided into two categories: conservative forces and nonconservative forces.

CONSERVATIVE FORCES

If the amount of work done in moving from one position to another position in a force field is independent of the path taken, the force is said to be *conservative.* Another test to see if a force is conservative or not is to apply it to an object through a closed path so you return to the starting point. If the amount of energy remains unchanged, the force is conservative. Examples of conservative forces are the forces of gravity and the spring force. For conservative forces we have $W_c = -\Delta PE$.

NONCONSERVATIVE FORCES

A nonconservative force is defined as a force that can change the mechanical energy (kinetic plus potential) of a system and its surroundings. Examples of nonconservative forces include the various kinds of friction encountered in motion. A force is nonconservative if it leads to a dissipation of mechanical energy. Because energy is lost, the path over which the force is applied will make a difference as to how much energy is lost and after returning to the starting point there will be less mechanical energy than before.

5.5 Conservation of Mechanical Energy

The sum of potential and kinetic energy, E, is called mechanical energy. If both KE and PE are present but no other forms of energy are, the sum, E, is also conserved. Thus for only mechanical energy present

$$KE_i + PE_i = KE_f + PE_f = E = \text{constant} \qquad (5.7)$$

Since E is constant, it does not change with time. Applying this to the gravitational force we get

$$\frac{1}{2}mv_i^2 + mgy_i = \frac{1}{2}mv_f^2 + mgy_f \qquad (5.8)$$

If there is more than one conservative force acting on a body then the total energy, E, is the sum of all the potential energies plus kinetic energy.

POTENTIAL ENERGY STORED IN A SPRING

The force of the spring is always opposite to the displacement, whereas the force we apply to the spring is aligned with the displacement. The force exerted by the spring is, therefore,

$$F_s = -kx \qquad (5.9)$$

The work we do on the spring in moving from 0 to some value of x is the average force $[\overline{F} = (0 + kx)/2 = kx/2]$ we apply times x and this work increases the potential energy of the spring so

$$W = \overline{F}x = \frac{1}{2}kx^2 = PE_s \qquad (5.10)$$

We define this energy as elastic potential energy.

5.6 Nonconservative Forces and the Work-Energy Theorem

Suppose W_{nc} represents the work down by nonconservative forces and W_c represents the work done by conservative forces, then according to the work-energy theorem, the net work on a system is the sum W_{nc} and W_c and is equal to the change in kinetic energy of the system. Also we know that $W_c = -\Delta PE$, so that

$$W_{nc} + W_c = \Delta KE \quad \text{so} \quad W_{nc} = \Delta KE - W_c$$
$$W_{nc} = \Delta KE + \Delta PE = E_f - E_i \qquad (5.11)$$

Thus we see that nonconservative forces are responsible for the change in total energy of the system. Frictional forces cause losses in mechanical energy. (Mechanical energy lost due to friction usually shows up as heat which is another form of energy. Thus, total energy, mechanical plus heat is still conserved.)

PROBLEM SOLVING STRATEGIES

See the text for some excellent problem solving strategies for solving problems involving mechanical energy and forces, both conservative and nonconservative. Also, carefully study the many excellent examples given in the text.

5.7 Conservation of Energy in General

On the microscopic level, that is, atomic level, the total energy is given by the sum of the kinetic energy and various potential energies. There are no examples observed to date of experiments wherein the total energy was not conserved. This statement includes energy in the form of mass with the energy equivalent of mass being given by Einstein's famous $E = mc^2$ equation. Thus we have the statement *"Energy can never be created or destroyed. Energy may be transformed from one form to another, but the total energy of an isolated system is always constant.* This implies that the total energy of the Universe is constant.

5.8 Power

Power is a measure of the rate in time at which work is done on system. The average power is the amount of work done divided by the time interval so

$$\overline{P} = \frac{W}{\Delta t} \tag{5.12}$$

Since work is $F\Delta s$ and $\Delta s/\Delta t$ is the average speed v we can also write

$$\overline{P} = \frac{W}{\Delta t} = \frac{\overline{F}\Delta s}{\Delta t} = \overline{F}v \tag{5.13}$$

The units of power is energy/time = J/s = watts = W. Another unit is horsepower which is given by 1 hp = 550 ft-lb/s = 746 W.

5.9 Work Done by a Varying Force

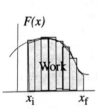

If the force varies with distance we have to figure the amount of work over very small displacements so the average force for that displacement can be taken and multiplied by Δx and the product then summed. Thus we get

$$W = \sum_{x_i}^{x_f} F_x \Delta x \tag{5.14}$$

This sum is equal to the area under the curve of F_x plotted versus x.

5.10 Concept Statements and Questions

1. In physics the work done by a force is defined as the product of the force and the displacement parallel to the force.

2. The component of force that is perpendicular to the displacement does zero work. Therefore, if a person carries a heavy weight horizontally, he does no work in moving it.

3. A horizontally applied force which acts on a free particle will change its kinetic energy by an amount that is equal to the net work done. What is the work-energy theorem? When does it apply?

4. What is a conservative force? How does it differ from a nonconservative force?

5. Potential energy is energy a particle has by virtue of its position. It can only be defined for conservative forces.

4. Power is the time rate of change of doing work.

5. The work-energy theorem is a consequence of conservation of energy. How can the principle of conservation of mechanical energy help us solve problems?

5.11 Hints for Solving the Problems

General Hints

1. Remember to take the component of the force along the displacement for calculating work. The perpendicular component does no work.

2. If work done on an object changes its kinetic energy, will the work necessarily increase the kinetic energy? Is kinetic energy always positive? How about the change in kinetic energy?

3. For power problems, if the velocity is known or can be easily calculated, then $\mathbf{F} \cdot \mathbf{v}$ is probably the easiest way to calculate the power.

4. This chapter has good examples for each of the topics treated. Make use of them.

Hints for Solving Selected Problems

1-3. Use the formula for work.

4-5. Take the component of force in the direction of the displacement or vice versa.

7. Use the formula for gravitational potential energy.

8. Use the work formula.

11. There is only horizontal velocity at the maximum height and it remains constant.

12-14. Apply the work-energy theorem to get work done and hence the average force times distance.

22 and 25. Use conservation of energy applied to the top and bottom of swing.

23, 24, 26-28. Use potential energy formula and remember that only the change in vertical height makes a change in potential energy.

29-34. It is important to watch the units carefully as you apply the work-energy theorem to these problems. You will note that the work-energy theorem applies to the work required to compress a spring and the energy stored in the spring. Use the kinetic energy formula for calculating kinetic energy when the velocity is given.

35-44. The loss in mechanical energy enables you to find the amount of work done by frictional forces.

45-54. Use the definition of power and the equations given. Convert between horsepower and watts as needed with the conversion factor given. For raising a weight along an incline, remember that only the vertical direction counts since the horizontal component is perpendicular to the gravitational force.

55-57. To get the work you need the area under the curve of F plotted against x. For straight lines it is easy to get the area from simply geometry.

6
Momentum and Collisions

Another Aspect of Inertia

Instead of $F = ma$, Newton's second law can be written as the time rate of change of momentum, which is defined as $p \equiv mv$. From Newton's third law, the force of action and the force of reaction are equal and opposite, so the two objects in collision experience equal and opposite forces which results in equal and opposite changes in momentum. These concepts enable us to better understand interactions and particularly collisions and explosions of objects.

6.1 Momentum and Impulse

The linear momentum, **p**, of a particle of mass m and velocity **v** is the simple product of the two, namely $\mathbf{p} = m\mathbf{v}$. Since this is a vector equation, then **p** has three components and $p_x = mv_x$, $p_y = mv_y$, and $p_z = mv_z$.

According to Newton's second law $F = \Delta p / \Delta t$ for each direction in space, so

$$\mathbf{p}_f - \mathbf{p}_i = \Delta \mathbf{p} = \mathbf{F}\Delta t = m\mathbf{v}_f - m\mathbf{v}_i \qquad \text{(6.1)}$$

The quantity Δp is called the impulse of the force. The force may be a complicated function of time, but its average times the time interval, Δt, is simply the impulse. Its magnitude is the area under the force versus time curve. The average force would be $\Delta p/\Delta t$, i.e., the impulse divided by the time difference or interval.

The impulse approximation is used in some problems. We assume that one of the forces acting for a short time on a particle is much greater than any of the other forces present.

6.2 Conservation of Momentum

If two particles collide, during the collision there is an equal and opposite force on each particle, so each will experience an equal but opposite change in momentum. The system, i.e., the combination of the two particles together, does not experience a change in momentum because $\mathbf{F}_{net}$ is zero. This is the basis for saying momentum is conserved. Internal forces cannot produce a net change in the momentum of the system. If there are n particles in the system we can write

$$\mathbf{p}_1 + \mathbf{p}_2 + \mathbf{p}_3 + \ldots + \mathbf{p}_n = \mathbf{P} = \text{constant} \qquad \text{(6.2)}$$

For collisions of two or more particles we see the total momentum, $\mathbf{P}$, before the collision is equal to the total momentum after.

6.3 Collisions

As we further consider the collision between two particles, we assume that the forces during the collision dominate all other forces present so only they need to be considered. The problem can be viewed as an impulse problem and the conservation of momentum is assumed to hold. The forces in the collision are equal and opposite so each object receives the same impulse and the same change in momentum, but opposite in direction. Thus $\mathbf{P}_{before} = \mathbf{P}_{after}$ where $\mathbf{P} = \mathbf{p}_1 + \mathbf{p}_2$.

An elastic collision is one in which momentum is conserved and also kinetic energy doesn't change. In inelastic collisions, momentum is still conserved, but due to nonconservative forces being present, the kinetic energy changes. For example, in a collision between two cars, if the fenders and bumpers are bent and crushed, then the mechanical energy that went into deforming them is lost. Momentum is constant for the colliding bodies unless there are external forces that are significant compared to the collisional forces.

PERFECTLY INELASTIC COLLISIONS

These are collisions wherein the two particles stick together and travel off with exactly the same velocity after the collision occurred. Thus in one dimension

$$m_1 v_{1i} + m_2 v_{2i} = (m_1 + m_2) v_f \tag{6.3}$$

ELASTIC COLLISIONS

Both momentum and kinetic energy are conserved. So

$$m_1 v_{1i} + m_2 v_{2i} = m_1 v_{1f} + m_2 v_{2f} \qquad \text{(6. 4)}$$

$$\frac{1}{2} m_1 v_{1i}^2 + \frac{1}{2} m_2 v_{2i}^2 = \frac{1}{2} m_1 v_{1f}^2 + \frac{1}{2} m_2 v_{2f}^2 \qquad \text{(6. 5)}$$

These two equations are solved together simultaneously. Various simplifications can be made and we get as solutions for the final velocities

$$v_{1f} = \frac{(m_1 - m_2)}{(m_1 + m_2)} v_{1i} + \frac{(2m_2)}{(m_1 + m_2)} v_{2i} \qquad \text{(6. 6)}$$

$$v_{2f} = \frac{(2m_1)}{(m_1 + m_2)} v_{1i} + \frac{(m_2 - m_1)}{(m_1 + m_2)} v_{2i} \qquad \text{(6. 7)}$$

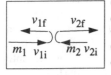

Other special cases can be easily handled from these equations. For example, if particle 2 is initially at rest, then we can drop the last term in each of the equations. Or if the two masses are equal, the terms involving the difference in masses drop out.

6.4 Glancing Collisions

Two dimensional collisions differ from one dimensional collisions in the fact that momentum equations are written for two dimensions instead of one. Some simplification is possible by choosing one of the axes to be colinear with the momentum of one of the incoming particles. Further simplification occurs also if one of the particles is initially at rest. For elastic collisions we still write down the conservation of kinetic energy, which being a scalar only gives one equation. Altogether, therefore, we end up with three equations that must be solved together. They are

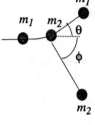

$$m_1 v_{1i} + 0 = m_1 v_{1f} \cos\theta + m_2 v_{2f} \cos\phi \qquad \text{(6. 8)}$$

$$0 + 0 = m_1 v_{1f} \sin\theta - m_2 v_{2f} \sin\phi \qquad \text{(6. 9)}$$

$$\frac{1}{2} m_1 v_{1i}^2 + 0 = \frac{1}{2} m_1 v_{1f}^2 + \frac{1}{2} m_2 v_{2f}^2 \qquad \text{(6. 10)}$$

If the two masses are equal, then $\theta + \phi = \dfrac{\pi}{2} = 90°$ **always.** Note θ is the angle that m_1 scatters into and ϕ is the angle at which m_2 scatters.

6.5 Concept Statements and Questions

1. Conservation of momentum can be applied when the net external forces are zero.

2. In collisions we assume the external forces are negligible, hence the momentum of the system before is equal to the momentum of the system after the collision.

3. How does impulse differ from momentum? You can determine impulse from the area under a force-time curve. What else?

4. How do elastic collisions differ from inelastic collisions?

5. In glancing collisions of two particles why are the particles always considered to move within a flat plane?

6.6 Hints for Solving the Problems

General Hints

1. Consider if the external forces are zero so you can apply conservation of momentum to your problem.

2. The c.m. is affected only by external forces.

3. Determine impulse from either a change in momentum or as the area under a force-time curve. The average force is determined by dividing the area by the time interval.

4. In problems like a person moving from one end of a canoe to the other end, use the concept that the c.o.m. did not move, so other parts of the system had to move to compensate.

5. In elastic collisions look for simplifying assumptions such as one of the particles being at rest or the fact that both masses are equal so the angle between the outgoing particles is $90°$.

Hints for Solving Selected Problems

1-4. These problems require applying the definition of momentum.

5. Rather than thinking about force as $\mathbf{F} = m(\Delta\mathbf{v}/\Delta t = a)$, think about $\mathbf{F} = (\Delta m/\Delta t)\mathbf{v}$.

6-8. Consider the force as $\Delta\mathbf{p}/\Delta t$.

9-13. Think about impulse being the area under a force $- \Delta t$ plot.

14. Think about combining mg with a (m/t)v term.

15-22. The impulse on each object is the same.

23-40. This is a huge set of problems involving collision. Several require only that momentum be conserved. The problems involving elastic collisions are easier after you carefully study the examples in the book. For inelastic collisions you can find the velocities before and after collision from the momenta and with these velocities you compute the kinetic energies, and then you can compute the change in kinetic energy.

7
Circular motion and the Law of Gravity

**A Study of the
Orbits of Planets
and Electrons**

Circular motion deals with angular displacements, angular velocities (speeds) angular and centripetal accelerations, and centripetal forces. The orbits of the planets are approximately, but not quite, circular. The force of gravity provides the necessary centripetal force.

7.1 Angular Speed and Angular Acceleration

The length of an arc, s, on a circle of radius, r, is related to r and the size of the angle θ making the arc according to

$$s = r\theta \qquad (7.1)$$

where θ is measured in radians and 2π radians $= 360^\circ$. A difference between two positions on the arc of a circle gives a displacement $\Delta s = s_2 - s_1$ and the difference between the two angles making up the arc, $\Delta\theta = \theta_2 - \theta_1$, is the angular displacement. The average angular speed $\overline{\omega}$ is given by

$$\overline{\omega} = \frac{\theta_2 - \theta_1}{t_2 - t_1} = \frac{\Delta\theta}{\Delta t}$$

(7.2)

The instantaneous angular speed is the limit of the average when the time interval, Δt, goes to zero.

7.2 Rotational Motion under Constant Angular Acceleration

By analogy with linear motion, we can define angular acceleration as

$$\overline{\alpha} \equiv \frac{\Delta\omega}{\Delta t} \quad \text{and} \quad \alpha_{instant} \equiv \lim_{\Delta t \to 0} \frac{\Delta\omega}{\Delta t}$$

(7.3)

We can then write down by analogy with linear motion under constant acceleration the kinematic equations for angular quantities when the angular acceleration is constant, namely

$$\omega = \omega_0 + \alpha t$$

$$\theta = \omega_0 + \frac{1}{2}\alpha t^2$$

$$\omega^2 = \omega_0^2 + 2\alpha\theta$$

(7.4)

7.3 Relations Between Angular and Linear Quantities

Since $s = r\theta$ we note the relationship between linear displacement and angular displacement is the radius, r. It follows that other angular variables are similarly related to the linear variable counterpart. Thus, we have for measurements along the arc (tangential variables)

$$s = r\theta \qquad \Delta s = r\Delta\theta$$
$$v_t = r\omega \qquad \Delta v_t = r\Delta\omega \qquad \text{(7.5)}$$
$$a_t = r\alpha$$

7.4 Centripetal Acceleration

We have uniform circular motion when there is an acceleration of constant magnitude continually occurs in a direction that is perpendicular to the velocity. This acceleration is called centripetal acceleration since it always points towards the center of the circle. It is along the radius of the circular motion it produces, but oppositely directed. It's magnitude is given by

$$a_c = \frac{v^2}{r} = \frac{(r\omega)^2}{r} = r\omega^2 \qquad \text{(7.6)}$$

If there is also tangential acceleration present the motion may still be circular but it is no longer uniform. The tangential and centripetal accelerations are vectors perpendicular to each other so the magnitude of the resultant acceleration may be obtained from the Pythagorean theorem:

$$a = \sqrt{a_t^2 + a_c^2}$$

(7. 7)

7.5 Centripetal Force

If a particle of mass m is undergoing centripetal acceleration, there must be a net force causing it and the force is called centripetal force. It is parallel to the acceleration and, therefore, along $-r$.

$$F_c = ma_c = \frac{mv_t^2}{r} = mr\omega^2$$

(7. 8)

7.6 Describing Motion of a Rotating System

In order to remain in circular motion there must be centripetal accelerations present. Thus an observer in a rotating reference frame is in an accelerated reference frame. Newton's laws as he gave them to us apply only to inertial (non-accelerating) reference frames. To pretend one is in an inertial frame when actually in a rotating one, requires that one introduces fictional forces to explain the motion of objects. Some of the fictitious forces are the "reaction" forces to the centripetal forces. These non-real forces, called centrifugal, seem real as long as the observer fails to recognize he is in a rotating frame of reference. The surface of the earth rotates about an axis through the poles. Therefore, we observe small centrifugal forces. The measured weight on a spring scale is a little less than it would be if the earth were not rotating.

7.7 Newton's Universal Law of Gravitation

Newton published his theory of gravitation in 1687 which states: **Every particle in the Universe attracts every other particle with a force that is directly proportional to the product of their masses and inversely proportional to the square of the distance between them.**

We have previously described the magnitude of this force as being "weight". The magnitude of the gravitational acceleration was g so $w = mg$. The gravitational acceleration depends on the mass of the object and the distance from it so

$$g = \frac{Gm_2}{r^2} \quad \text{so that} \quad F_g = \frac{Gm_1m_2}{r^2} = m_1g \quad \text{(7.9)}$$

where G is the *universal gravitational constant,* with the value $G = 6.67 \times 10^{-11}$ N. m^2/kg^2. When $r = R_E$, the radius of the Earth, and $m_2 = M_E$, the mass of the Earth, the value of g is 9.8 m/s^2 and F_g equals the weight of the object at the Earth's surface.

7.8 Kepler's Laws

Using the data about the positions of the planets in the sky as measured by Tycho Brahe, Johannes Kepler, an assistant of Brahe, deduced three laws of planetary motion. They are:

1. Every planet moves in an elliptical orbit with the Sun at one of the focal points.

2. The radius vector drawn from the Sun to any planet sweeps out equal areas in equal time intervals.

3. The square of the orbital period of any planet is proportional to the cube of the semimajor axis of the elliptical orbit (the average distance of the planet from the sun as it completes an orbit).

Newton was able to use his three laws of motion along with his law of gravity and derive these three laws of Kepler. Newton was able to derive the constant of proportionality in Kepler's third law and from this it is possible to get the masses of the moon, sun, planets and various satellites. This was done as follows.

KEPLER'S THIRD LAW

By equating the gravitational force of the sun on a planet to the centripetal force (assuming a circular orbit) and setting $v = 2\pi r/T$ (T is the period of revolution) one readily derives Kepler's third law and obtains a value for Kepler's proportionality constant from which the mass of the sun can be calculated.

$$\frac{GM_sM_p}{r^2} = \frac{M_pv^2}{r}, \text{ so } T^2 = \left(\frac{4\pi^2}{GM_s}\right)r^3 = K_sr^3, \text{ where}$$

the value of K_s is, $K_s = 2.97 \times 10^{-19} \text{ s}^2/\text{m}^3$.

These equations also apply to planets with a somewhat different value for K_s since the mass of the planet is different than the sun's mass. By observing the satellites orbiting a planet, we can use Kepler's third law to get the mass of the planet.

7.9 Concepts Statements and Questions

1. The equations of rotational motion are similar to those of linear motion. We work with rotational variables instead of linear ones. Otherwise, the equations are completely analogous.

2. How does one transfer over from the linear kinematic variables to the angular variables and vice versa?

3. What is centripetal force? Is it the same as centrifugal force? (They are numerically equal.) So why do we talk about them both? We should avoid using language involving fictitious forces as much as possible.

4. Since the centripetal force is perpendicular to the motion of the object as it travels in a circular path, how much work does it do? Therefore, how much work does the earth do on the moon each month as it travels about the earth?

5. Does the weight of a person vary with height as she moves out from the earth? Explain.

6. Since Newton can derive Kepler's laws from his laws of motion and his gravitational law, we recognize Newton's laws as being more fundamental.

7.10 Hints for Solving the Problems

General Hints

1. Study all of the examples given in the text very carefully.

2. Practice going from rotational variables and equations to linear ones, like what is given in Example 7.2

3. Follow the problem-solving strategy for centripetal forces given in the text in Section 7.5.

Hints for Solving Selected Problems

1-5. Use $s = r\theta$ and convert between degrees and radians by π radians $= 180^\circ$.

6-8. In addition to the hints for 1-5. remember that $\theta = \omega t$.

9-17. These problems are just like working with the kinematic equations in chapter 3, but using angle variables instead.

18-28. In this centripetal acceleration problems we use the hints thus far but include the equations for centripetal acceleration in terms of $a_c = v^2/r = r\omega^2$.

29-40. From the centripetal acceleration you can get the centripetal force from Newton's second law, $F = ma$. This force could be a tension in a rope, the force on a spring or even a frictional force. If velocity changes, the force will probably have to change also.

41-52. Remember the gravitational force is a vector and if two or more are added together, the addition rules for vectors must be followed.

8
Rotational Equilibrium and Dynamics

Motion About a Fixed Axis

Objects at rest or moving at constant velocity are said to be in equilibrium. Buildings, bridges, and biological systems at rest are all examples of the application of equilibrium concepts. From chapter 7 we already have the definitions we need for rotational kinematic equations. From there we can get the dynamical equations, talk about torque (analogous to force), angular momentum (analogous to linear momentum), rotational kinetic energy, etc.

8.1 Torque

The ability of a force to rotate a body about some axis is measured by a quantity called torque, τ. Torque, like force, is a vector. Its magnitude is

$$\text{Torque} \qquad \tau = Fd \qquad \text{(8.1)}$$

d is the lever arm and is the perpendicular distance from the axis of rotation to a line drawn along the direction of force.

If there are several forces acting on the object then the net torque is obtained by summing the torques produced by each of the forces, thus

$$\tau_{net} = \sum \tau = \tau_1 + \tau_2 + \dots = F_1 d_1 + F_2 d_2 + \dots \qquad \text{(8.2)}$$

8.2 Torque and the Second Condition for Equilibrium

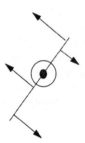

In section 4.6 we discussed the first condition for equilibrium, namely the sum of all the external forces is zero. With no net forces acting on the object it obeys Newton's first law, i.e. no accelerations and thus no changes in motion. ($\sum F = 0$) The object could still rotate and even change its rate of rotation if the torques don't add to zero.

The second condition for equilibrium, therefore, is that there are no changes in rotation. This happens when the sum of the external torques adds to zero, so we have

$$\sum \tau = 0 \qquad \text{(8.3)}$$

POSITION OF THE AXIS OF ROTATION

If the object is in equilibrium, it does not matter where you put the axis of rotation for calculating the net torque; the location of the axis is completely arbitrary.

8.3 The Center of Gravity

The center of gravity is that point in or near an object where all the torques due to the weight of the object add to zero no matter how the object is oriented. It is what we usually call the balance point. The x component is found from the equation

$$x_{cg} = \frac{m_1 x_1 + m_2 x_2 + \dots}{m_1 + m_2 + \dots} = \frac{\sum_i m_i x_i}{M} \qquad \text{(8. 4)}$$

The y and z components of the center of gravity are found in a completely analogous manner.

The center of gravity of a symmetrical body that is homogenous must lie on the axes of symmetry.

8.4 Examples of Objects in Equilibrium

For objects in equilibrium both the net force and the net torque must add to zero. The text gives many excellent examples of solving various problems involving equilibrium. Also the text gives a section on problem-solving strategies. Study the strategies and examples carefully.

8.5 Relationship Between Torque and Angular Acceleration

The angular acceleration is proportional to the applied torque, thus

$$\tau = rF = rma = mr^2\alpha \tag{8.5}$$

Here we used the relationship that $a = r\alpha$ and have taken r to represent the perpendicular distance from the point to the axis.

TORQUE ON A ROTATING OBJECT

For an extended object such as a disk we find mass distributed at various distances from the axis, so we must sum up all of these masses for each r, and of course sum up all of the torques applied to give a total torque, thus,

$$\sum \tau = \left(\sum mr^2\right)\alpha = I\alpha \tag{8.6}$$

The variable I is called the moment of inertia and has units of kg·m². It plays a role for rotational motion similar to mass in linear motion. We see that $\tau = I\alpha$ plays a role for rotation similar to $F = ma$ for linear motion.

8.6 Rotational Kinetic Energy

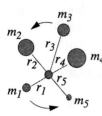

A rigid body undergoing pure rotation can be considered to be made up of many little masses, m_i, each having a speed v_i, so that the total kinetic energy is

$$KE = \frac{1}{2}\sum m_i v_i^2 = \frac{1}{2}\sum \left(m_i r_i^2\right)\omega^2 = \frac{1}{2}I\omega^2 \tag{8.7}$$

Here we have introduced the variable I which is the moment of inertia, $I = \sum m_i r_i^2$.

8.7 Angular Momentum

The angular momentum of an object with mass, m, revolving around an axis is given as $L = mrv = mr^2\omega$. Thus for a rotating object the angular momentum can be given as

$$L \equiv I\omega \qquad \text{(8. 8)}$$

where ω is the angular speed and I is the moment of inertia.

8.8 Conservation of Angular Momentum

Linear momentum is conserved when the net external force is zero. Since the net external torque is equal to $I\alpha$ and α is the time rate of change of the angular velocity, $\Delta\omega/\Delta t$, we see that *torque*, τ, must be the time rate of change of angular momentum L. Whenever the external torque is zero, L must be constant, i.e.

$$L_{before} = L_{after} \qquad \sum \tau_{ext} = \frac{\Delta \mathbf{L}}{\Delta t} = 0 \quad \text{implies that} \quad \mathbf{L} = cons\tan t \qquad \text{(8. 9)}$$

Thus for an isolated system angular momentum is conserved just like linear momentum and kinetic energy are.

Since $L_f = L_i$, we know

$$L = I\omega \quad \text{so} \quad I_f\omega_f = I_i\omega_i \qquad \text{(8. 10)}$$

Thus, in an isolated, rotating system I and ω may individually change in time but the product, $I\omega$, will not.

8.9 Quantization of Angular Momentum

On the atomic level energy can be quantized. Also angular momentum comes in quantized amounts with $h/2\pi$ being the fundamental unit with a value of 1.054 x 10^{-34} kg-m^2/s.

Many of the fundamental particles possess an intrinsic amount of angular momentum known as the "spin" of the particle. Protons, neutrons, and electrons each have 1/2 of the fundamental unit, whereas the photon (the particle of light) carries one whole unit. Also orbital angular momentum of the electrons in atoms have integer amounts of the fundamental unit.

8.10 Concept Statements and Questions

1. Every kinematic and dynamic equation we developed in non-rotational motion has an analog in rotational motion. The equations differ mostly by a multiplicative factor of r or r^2, except for those involving the moment of inertia, which essentially swallowed an r^2 in its definition.

2. Torque is a force acting through a perpendicular lever to the rotation axis. It is the analog of force for rotational motion and causes angular accelerations.

3. To be in equilibrium all of the external forces must add to zero and for rotational equilibrium, all of the external torques must add to zero.

4. In order to balance, where does the center of gravity have to be with respect to the point of support?

5. Force causes momentum (mv) to change. What causes angular momentum ($I\omega$) to change?

6. Angular momentum is conserved when the sum of the external torques adds to zero. I and ω can change but the product will not if angular momentum is conserved. It is an intrinsic property on the atomic level of elementary particles.

8.11 Hints for Solving the Problems

General Hints

1. In adapting kinematic and dynamic equations to rotational variables one must check the units carefully and be sure not to pick up or drop factors of r. List and study the equations to become familiar with them. They will be used frequently in the chapters that follow.

2. The center of gravity does not move if the sum of the external forces are zero.

3. Become familiar with the various equations for angular momentum and how they relate to each other. Why, for example, does the angular velocity increase when r is decreased? What about linear velocity? Does it also increase? Which equations will help you decide.

4. Revolutions per minute or per second (rpm or rps) can be converted to radians per minute or second by multiplying by 2π.

Hints for Solving Selected Problems

1-6. These are practice problems for learning the definition of torque. Use the definition equation.

7-23. These problems can be a little tricky so it would be well to study the examples in the text very carefully. The key to these problems is balancing torques around a point so there are no rotations while also making sure the external forces add to zero so there are no accelerations.

24-39. The net torque in these problems produces a change in the angular speed. These problems are like, $F = ma$ problems. To get angular acceleration use angular kinematic equations. Be sure you understand how to get the moment of inertia for various simple situations.

40-52. Rotational kinetic energy is obtained by getting the moment of inertia and the angular speed. You need to get these variables into the right units. Angular speed should be in radians per second in the rotational kinetic energy formula.

53-62. Angular momentum is as easy to work with as linear momentum. It is conserved if there are no net external torques. Nevertheless, if moment of inertia changes, angular speed will have to change as well. In Problem 62, the woman provides the energy, but still there are no external toques, so angular momentum is conserved.

9
Solids and Fluids

**Behavior and
Properties of
Matter in Bulk
Quantities**

Matter in bulk comes as solids and fluids (both liquid
and gas fluids). An understanding of fluids is particularly important to students going into the life sciences.

9.1 States of Matter

Matter is found in four states or conditions, the three
most common being solid, liquid and gas. The fourth
state occurs when some of the electrons are physically
separated from the rest of the material leaving it ionized. Such a state of matter is called a plasma.

Solids are held together by bonds and can be classified
as either crystalline (ordered arrangement of the
atoms) or amorphous (random arrangement). When the
intermolecular forces cannot hold the atoms into any
position we have a fluid, and when they are so closely

packed that they cannot be easily compressed, we have a liquid. In a gas the molecules move about freely and large distances exist between the particles, so a gas can be readily compressed.

9.2 The Deformation of Solids

There is no perfectly rigid material, hence all solids can be deformed (stretched, compressed, sheared, etc.) to some extent. The elastic properties can be discussed in terms of stress (deforming force per unit area) and strain, a measure of the degree of deformation. Stress is proportional to strain up to a point, whereupon the elastic limit is exceeded and the deformation becomes permanent. The constant of proportionality is called the elastic modulus, so

$$\text{Elastic Modulus} = \frac{stress}{strain} \tag{9.1}$$

YOUNG'S MODULUS: ELASTICITY IN LENGTH

The stress is force/area and is measured in N/m^2 or pascals. In stretching a wire, the elastic modulus is called Young's modulus, Y, and is defined by

$$Y \equiv \frac{\text{tensil stress}}{\text{tensil strain}} = \frac{F/A}{\Delta L/L_0} = \frac{FL_0}{A\Delta L} \tag{9.2}$$

SHEAR MODULUS: ELASTICITY OF SHAPE

When the force is tangential to the face while the other face is held fixed we get shear deformation and shear modulus given by

$$S \equiv \frac{\text{shear stress}}{\text{shear strain}} = \frac{F/A}{\Delta x/h} \qquad \text{(9.3)}$$

where Δx is the shear displacement and h is the height of the object.

BULK MODULUS: VOLUME ELASTICITY

This applies to volume compression or expansion. The reciprocal of the bulk modulus, B, is called the compressibility. B is defined as

$$B \equiv \frac{\text{volume stress}}{\text{volume strain}} = -\frac{F/A}{\Delta V/V} \qquad \text{(9.4)}$$

Note the minus sign. The volume gets smaller as the compressional force gets larger.

9.3 Density and Pressure

The density is defined as mass per unit volume. It is represented by ρ, thus

$$\rho = \frac{M}{V} \qquad \text{(9.5)}$$

The specific gravity is the ratio of the density of a substance to the density of water at $4°C$, which is 10^3 kg/m^3. Pressure, like stress, is force/unit area with units of pascal or Pa, so

$$P \equiv \frac{F}{A} \qquad \text{(9.6)}$$

9.4 Variation of Pressure with Depth

The weight due to a column of material exerts a pressure at the bottom of the column equal to the weight spread over the area, or $P = w_g/A$. Since the mass is the density times the volume (true for constant density such as a solid or liquid), $w_g = mg = \rho g V = \rho g A h$, and we see the pressure $P = \rho g h$. The total pressure at the bottom of a column is $\rho g h$ plus the pressure at the top P_0, so

$$P = P_0 + \rho g h \qquad \text{(9.7)}$$

Pascal (1623-1662) gave us Pascal's principle which is: *Pressure applied to an enclosed fluid is transmitted equally in all directions to every point of the fluid and the walls of the containing vessel.*

9.5 Pressure Measurements

The open tube manometer is frequently used to measure pressures. It is a U-shaped, vertical tube containing a liquid open to the atmosphere at pressure P_0 on one end, and the other end is connected to a system of unknown pressure P. The pressure at a point B below the surface is $P_0 + \rho g h$, where ρ is the density and h is the height of the fluid. If the fluid on the other end is at the same level as B, then the unknown pressure is also $P_0 + \rho g h$.

$P_0 + \rho g h$ is the absolute pressure and $P - P_0$ is called the gauge pressure.

The mercury barometer is another instrument used to measure pressure. It is made of an inverted tube filled with mercury and the open, bottom end is inserted into a container of mercury. the top end is closed. The mercury rises to a height, h, so that $P_0 = \rho g h$.

One atmosphere of pressure is defined to be the equivalent pressure produced by a column of mercury 76 cm high and is equal to 1.013×10^5 Pa.

9.6 Buoyant Forces and Archimedes' Principle

$F_{buoy} = B$

Archimedes, the ancient Greek mathematician, discovered an important principle about objects in fluids, namely, *any body completely or partially submerged in a static fluid is buoyed up by a force equal to the weight of the fluid displaced by the body.*

We now understand the buoyant force to arise from the difference between the force due to pressure on the underside pushing up and the downward for due to the pressures on top.

$w = mg$

You can also think of it as follows. When placed in a fluid the object displaces a certain volume of fluid equal to the volume of the part of the object submerged, and it requires a force to lift this fluid up and out of the way. This is equal to the buoyant force, B.

$$B = \rho_{fluid} V_{submerged} g \qquad \text{(9.8)}$$

9.7 Fluids in Motion

Fluid in motion is said to be *streamline (laminar)* if every point moves along a smooth path or *turbulent* if it doesn't. Streamlines will not cross each other, whereas turbulent flow is irregular with swirls and eddy currents.

The ideal fluid is non-viscous (no internal frictional forces between streamlines), incompressible, steady, and laminar.

EQUATION OF CONTINUITY

In order to have conservation of matter in the flow of an ideal fluid, the amount of fluid passing one point per unit time must equal the amount of fluid passing another point. This gives the equation $\rho A_1 v_1 = \rho A_2 v_2$

or by canceling out the densities,

$$A_1 v_1 = A_2 v_2 \qquad (9.9)$$

which is known as *the equation of continuity.*

BERNOULLI'S EQUATION

Applying Newton's second law and Pascal's principle to fluids we get Bernoulli's equation, namely

$$P_1 + \frac{1}{2}\rho v_1^2 + \rho g y_1 = P_2 + \frac{1}{2}\rho v_2^2 + \rho g y_2$$

$$\text{or} \qquad P + \frac{1}{2}\rho v^2 + \rho g y = \text{constant} \qquad (9.10)$$

Note that if the flow is horizontal so that $y_1 = y_2$, then the pressure is lowest where the speed is greatest and from the equation of continuity this is where the area is smallest.

9.8 Other Applications of Bernoulli's Equation

Bernoulli's equation readily explains many interesting and curious phenomena and has many practical applications as well. The text explains several effects which we simply list here without detailed explanation.

The curve ball in baseball, softball, tennis, golf, etc. Due to the spin on the ball, the speed of air on one side is less than on the other and hence there is a difference in pressure.

The venturi tube is a horizontal pipe which has a smaller cross sectional area in the center so the speed of the fluid is higher and the pressure is reduced in that region.

A stream of gas flowing over an open tube reduces the pressure in the tube. Wind blowing over the top of a chimney provides a draft for the fire below.

Vascular flutter produced in a person with advanced arteriosclerosis is due to increased speed of blood flow through restricted veins. Also in an aneurysm, which is a weakened artery where the walls have ballooned outwards, the increased area results in slower flow and thus increased pressure.

The lift on an airplane wing comes about because the wing is shaped so the flow over the top of the wing is faster than over the bottom, so the pressure on top is less.

9.9 Concept Statements and Questions

1. What are the four states of matter?

2. As a material is heated, the motions of the atoms or molecules become more rapid so the bonds begin to weaken or break. How can this explaining melting and evaporation of a substance which begins as a solid?

3. When a solid is stretched too far it becomes permanently deformed. Up until the elastic limit is exceeded, most materials will go back to their original shape. How can this be understood from the idea of bonding between atoms?

4. The three most common types of deformation are: elongation or compression and characterized by Young's modulus; shear characterized by shear modulus, and volume change characterized by bulk modulus.

5. How is density related to mass and volume?

6. How does the weight of a column of fluid produce a pressure at the bottom of the column?

7. Distinguish between absolute pressure and gauge pressure.

8. Since pressure is a force per unit area. it is the same in all directions in a static fluid depending only on depth. This is called Pascal's principle.

9. The equation of continuity is possible because of what conservation principle?

9.10 Hints for Solving the Problems

General Hints

1. Be sure you understand the meaning of stress and strain and they are defined and used in each of the modulus equations.

2. Pascal's principle as applied to pistons can give a mechanical advantage. Be sure you are family with the principle and can apply it to static fluid situations.

3. Think about Archimedes' principle and how it relates to buoyant forces, and also in terms of the weight of the fluid displaced. Study the principle until you feel comfortable with it.

4. Study the examples given.

Hints to Solving Selected Problems

1-7. Use the equation for Young's modulus in variations of the formula.

8. Use equation for bulk modulus.

11. The shear area is the circumference times the thickness.

14-17. Use variations of the $F = P/A$ equation for various kinds of forces.

18-27. These problems involve heavily the idea of pressure due to a column of liquid given by ρgh. If there are several liquids, you need to determine the pressure for each and add them together.

28-39. You will use Archimedes' principle in a variety of ways in these problems. Buoyant forces arise from pressure differences. They also arise from fluid weight displaced. Once you compute the buoyant force, you can calculate the acceleration of an object subject to such forces. Kinematic equations apply for constant forces and accelerations.

40-44. In these problems you need to use the continuity equation.

45-57. You will need to apply Bernoulli's equation to most of these problems. Find the pressure difference in two places by comparing the speed of flow of the fluid and/or the elevation.

10
Thermal Physics

Large Numbers of Particles
Mechanics

Up till now we have been dealing with the motions of one or two objects, like in collisions. When we deal with large numbers of particles it is impossible to calculate collisions and trajectories and keep track of each one of the particles. So the properties of matter in bulk (pressure, temperature, entropy, and etc.) have to be learned and their relationships to each other established and also related to the variables (momentum, energy, etc.) which we have been studying up till now. We begin in this chapter by defining pressure and temperature and how materials behave as temperature changes.

10.1 Temperature and the Zeroth Law of Thermodynamics

Temperature is a measure of the average kinetic energy of all the particles in an object. When the energy is distributed uniformly we have equilibrium and the same temperature everywhere in the sample. If the energy is not uniform then non-equilibrium exists and some places are higher (or lower) in temperature than others. Two objects at different temperatures brought into contact with one another will exchange energy until equilibrium is achieved and they both have the same temperature. This means the average kinetic energy of each object is now the same.

The zeroth law of thermodynamics states: *If bodies A and B are separately in thermal equilibrium with a third body C, then A and B will be in thermal equilibrium with each other if placed in thermal contact.*

10.2 Thermometers and Temperature Scales

Thermometers are instruments designed to measure temperatures of objects. They generally work on the principle of change in a material that occurs due to a temperature change, e.g. volume, area, length, electrical resistance, color, etc.

Two commonly used temperature scales are Celsius and Fahrenheit. The Celsius (formerly called centigrade scale) is based upon 100 degree units between the freezing point and the boiling point of water at sea level. The Fahrenheit scale assumes there are 180 degrees between those two points and the freezing point of water is at $+32^\circ$F (so boiling is at 212°F).

The Constant-Volume Gas Thermometer and the Kelvin Scale

A pressure of a gas changes linearly with the temperature providing the volume is kept constant changes. Extrapolating the result to zero pressure, one reaches what is defined as absolute zero. This occurs for all gases at -273.15°C, but most cases liquify before reaching that temperature. On the Kelvin scale this is taken as 0 K and has gradations the same as Celsius, so water freezes (melts) at 273.15 K and it boils 100 degrees above that or 373.15 K. The kinetic energy of atoms and molecules is zero at 0 K.

The triple point of water where ice, water, and vapor can co-exist is at 0.01°C or 273.16 K.

The Celsius, Fahrenheit, and Kelvin Temperature Scales.

To change from one scale to the other the convenient formulas to use are

$$T_K = T_C + 273.15, \quad \text{and} \quad T_F = \frac{9}{5}T_C + 32 \qquad \text{(10.1)}$$

10.3 Thermal Expansion of Solids and Liquids

Most solids expand in length, area, and volume as temperature increases. This can be understood as an increase in the amplitude of vibration of the atoms or molecules of the material about their positions of equilibrium. The relationship is usually linear with temperature so a change in L is given by

$$\Delta L = \alpha L_o \Delta T, \quad \text{or} \quad L = L_o (1 + \alpha (T - T_o)) \quad \text{(10.2)}$$

The quantity α is called the coefficient of linear expansion and depends on the material.

For the increase in area and volume we have linear expansion in two and three dimensions respectively, so we can get the equation for expansion by squaring the equation for length to get area, and cubing the length to get volume. By keeping only the first order terms we get

$$\Delta A = 2\alpha A_o \Delta T, \quad \text{and} \quad \Delta V = 3\alpha V_o \Delta T$$
$$or$$
$$\Delta A = \gamma A_o \Delta T, \quad \text{and} \quad \Delta V = \beta V_o \Delta T \quad \text{(10.3)}$$

where $\gamma = 2\alpha$ and $\beta = 3\alpha$.

The Unusual Behavior of Water

The density of ice is less than the density of water. Also when liquid water goes from $0^\circ C$ to $4^\circ C$ its volume decreases, so it becomes more dense and sinks. Above $4^\circ C$ water expands like other liquids and the density decreases normally.

10.4 Macroscopic Description of an Ideal Gas

At very low pressure (or low density), gas at room temperature approximates what is called an ideal gas. An *equation of state* can be written which relates pressure, temperature, amount of gas present, and volume. The equation of state for an ideal gas is

$$PV = nRT \qquad \text{(10.4)}$$

where n is the number of moles of gas ($n = m/M =$ mass/molecular weight) and R is a proportionality constant called the universal gas constant. In SI units

$$R = 8.31 \text{ J/mol-K} = 0.0821 \text{ l-atm/mol-K} \qquad \text{(10.5)}$$

10.5 Avogadro's Number and the Ideal Gas Law

Based on experiments of dissociated liquids into their component gases, Avogadro was able to explain the results by assuming the volume of gas of any substance always contained the same number of molecules at the same temperature and pressure. We now know a mole of gas (a molar mass of the substance) contains $N_A = 6.02 \times 10^{23}$ molecules (Avogadro's number) and a mole of gas at standard temperature and pressure occupies 22.4 liters.

The ideal gas law can now be expressed in terms of the total number of molecules, N, and Boltzmann's constant, $k_B = R/N_A = 1.38 \times 10^{-23}$ J/K. Expressed in these variables the law becomes,

$$PV = nRT = \frac{N}{N_A}RT = Nk_BT \qquad \text{(10.6)}$$

10.6 The Kinetic Theory of Gases

Here we try to understand the equation of state and what is happening to the gas by looking at the gas on the atomic or molecular level.

MOLECULAR MODEL FOR THE PRESSURE OF AN IDEAL GAS

We assume

1. The number of molecules is large and the distances of separation are large compared to the size of the molecules.

2. The molecules obey Newtonian mechanics, but as a whole they move randomly.

3. The molecules undergo collisions that are on the average elastic.

4. The forces on the molecules are negligible except during collisions.

5. All of the molecules are the same.

The change in momentum in collision with the wall perpendicular to the x-axis is given by $\Delta p_x = mv_f - mv_i = -2mv_x$ and thus this is the impulse $(\overline{F}\Delta t)$ given to the wall when the molecule hits it. The frequency at which this occurs is $v_x/2d$, so the average force on the wall for each molecule is $m\overline{v}_x^2/d$. The average force due to N molecules would be N times this average force for each.

Because $\overline{v}^2 = \overline{v}_x^2 + \overline{v}_y^2 + \overline{v}_z^2$ and $\overline{v}_x^2 = \overline{v}_y^2 = \overline{v}_y^2$, it follows that $\overline{v}_x^2 = \overline{v^2}/3$ and the net force and pressure are given by

$$\overline{F}_{net} = \frac{N}{3}\left(\frac{m\overline{v}^2}{d}\right) \quad \text{and} \quad P = \frac{\overline{F}_{net}}{A} = \frac{2}{3}\left(\frac{N}{V}\right)\left(\frac{1}{2}m\overline{v}^2\right) \quad \text{(10.7)}$$

MOLECULAR INTERPRETATION OF TEMPERATURE

Comparing Equation 10.6 with the ideal gas law shows that temperature is related to average kinetic energy according to

$$\overline{KE} = \frac{1}{2}m\bar{v}^2 = \frac{3}{2}k_B T \qquad (10.8)$$

Each degree of freedom (*in this situation* x,y,z) contributes $\overline{KE}/3$ per molecule to the energy of the system. This statement is known as *the equipartition theorem*. The square root of the average velocity squared, v_{av}^2, is called the root mean square velocity and relates to T as

$$v_{rms} = \sqrt{v_{av}^2} = \sqrt{\frac{3k_B T}{m}} = \sqrt{\frac{3RT}{M}} \qquad (10.9)$$

10.7 Concept Statements and Questions

1. Pressure due to a gas is the net impulse per unit time delivered by the molecules of gas in the container per unit area.

2. The total kinetic energy contained in a gas is the heat and is the sum of the kinetic energy of all the molecules. (If the molecules rotate and/or vibrate, these forms of energy also contribute to the heat. So each degree of freedom contributes $(1/2)k_B T$ of energy per molecule to the system. This is according to the equipartition theorem.)

3. Temperature is a measure of the average kinetic energy of all the molecules in the substance (for temperatures that are not too high and not too low).

4. A temperature of absolute zero is reached when the average kinetic energy of the molecules is zero. (Actually the molecules never quite reach zero kinetic energy because of reasons we can't go into here.)

5. The pressure of an ideal gas at constant volume linearly with temperature.

6. The increase of length, area, and volume of a solid is due to an increased amplitude of vibration of the atoms or molecules making up the solid as the average energy ($\alpha\ T$) increases. How does the coefficient of expansion of an area or volume relate to the coefficient of linear expansion?

7. The ideal gas law $PV = nRT$ works well for most gases. Express the law in terms of the number of molecules instead of the number of moles.

10.8 Hints for Solving the Problems

General Hints

1. When converting Celsius to Fahrenheit and vice-versa, you will want to be careful about including the $32°F$ for the freezing point of water.

2. Be careful in your use of Nk_B as opposed to nR in working with the equation of state. Make sure you don't mix them up.

3. Be sure you understand the microscopic theory of gases and how it relates to the macroscopic theory.

4. When working with pressure in atmospheres, change to pascals before trying to give force in Newtons.

Hints for Solving Selected Problems

1-2. For the constant volume thermometer you can set $P = aT + b$, set up two such equations with P and T known and solve them simultaneously to get a and b. Once a and b are known you can get P for any temperature.

3-8. Convert temperatures using the conversion formula.

9. Use the conversion formula between Celsius and Fahrenheit and require $T_C = T_F$.

11-24. Use the equations for linear, area and volume expansion. Also remember that the coefficient of areal expansion is 2α and volume expansion is 3α, where α is the coefficient of linear expansion. Holes in solids expand the same as the material itself would expand.

25-37. These problems require use of the ideal gas law. For changes in pressure, ΔP you can use $V\Delta P = nR\Delta T$

45-46. Use the concept of impulse to get the force.

50-52. Use the temperature to get the average speed of the gas molecules and compare these to the escape speed from a planet.

11
Heat

Heat: Thermal Energy Transfer

Heat was once considered to be a mystical fluid called caloric, that could be made to flow from one substance to another. It is now recognized to be energy, and is specifically defined as energy transfer due to a teimperature difference.

The first law of thermodynamics is the law of conservation of energy and as such includes all forms of energy including thermal energy.

11.1 The Mechanical Equivalent of Heat

Heat is used to denote thermal energy that can be transferred from one substance to another when there is a temperature difference and they are in thermal contact. Historically this form of energy was denoted by units of calories and BTUs (British Thermal Units). These units have an equivalence to other energy units.

The calorie used to be defined as the heat necessary to raise the temperature of 1 gram of water from 14.5°C to 15.5°C. The kilocalorie (Calorie, spelled with capital C and used in counting food calories) is 1000 cal and will change the temperature of 1 kg of water from 14.5°C to 15.5°C. The cal is now defined in terms of mechanical energy in the next paragraph.

JOULE'S EXPERIMENT

Joule found he could increase the temperature of water by rotating a paddle wheel inside the water. He was able to measure the energy put into the paddle wheel in terms of joules and from the temperature rise, he determined the number of calories required to achieve the same rise and thus, he found the mechanical equivalent of heat. The number used to define the calorie exactly is

$$1 \text{ calorie} = 4.186 \text{ J} \qquad \text{(11. 1)}$$

(1 cal is defined in some books as exactly 4.184 J.)

11.2 Specific Heat

The amount of heat transfer required to raise the temperature of one gram of material by one degree Celsius is called the *specific heat* and the value of this specific heat varies from one substance to another. The specific heat of a substance is defined as

$$c = \frac{Q}{m\Delta T}, \qquad Q \text{ (heat = thermal energy transfer)} \quad \text{(11. 2)}$$

Specific heat at constant volume is different from specific heat at constant pressure. (Note: similar definitions are made in the British system of units, with the amount of heat required to raise one pound of water through one degree Fahrenheit being defined as 1 BTU, so 1 BTU = 252 cal.

Water has one of the highest specific heats among common materials.

11.3 Conservation of Energy: Calorimetry

We use conservation of energy to figure the temperature change when two materials of different temperature are brought together and allowed to come to equilibrium. One object loses energy and the other gains the same amount, so the net energy remains the same. Also if mechanical energy enters or leaves the system, this will cause a change in temperature. Keeping track of the energy lost and gained and the temperature changes that occur is called *calorimetry.*

11.4 Latent Heat and Phase Changes

When a substance changes phase, say from a solid to a liquid, a lot of energy is required. The heat required is called *latent heat (L).* When the substance melts the heat is called the heat of fusion and when it vaporizes, the latent heat involved is called heat of vaporization.

The heat for a given mass to change phase is

$$Q = mL \qquad \text{(11.3)}$$

The latent heat of fusion and of vaporization varies considerably from substance to substance.

11.5 Heat Transfer by Conduction

Thermal energy can be transferred by one of three methods, namely conduction, convection and radiation. While some conduction can occur in any material, good electrical conductors like the metals are also good heat conductors. The free electrons in the metal which conduct the electricity also transport thermal energy from a higher temperature location to a lower temperature position through diffusion. Without the free electrons to transport the energy, conduction occurs from atom to atom through increased amplitudes of vibration. Poor conducting materials are things such as asbestos, cork, paper, fiber glass etc., but diamond is one of the best conductors.

If Q is the amount of thermal energy transferred from one location on an object to another in the time Δt, the transfer rate, H, is defined as

$$H \equiv \frac{Q}{\Delta t} \tag{11.4}$$

and this is proportional to the area and to the difference in temperature and inversely proportional to the thickness, L, of the slab, so that

$$H = \frac{Q}{\Delta t} = kA \left(\frac{T_2 - T_1}{L} \right) \tag{11.5}$$

Different materials have different values for k.

HOME INSULATION

For compound slabs such as is found in buildings it is found that thermal energy transfer occurs according to

$$H = \frac{Q}{\Delta t} = A \left(\frac{T_2 - T_1}{\sum_i L_i / k_i} \right) \tag{11.6}$$

L/k is called the R value and for low conduction (good insulation), needs to be as large as possible within reason. Note, one sums the R values in the denominator.

11.6 Convection

Convection is an energy transfer process in which the material wherein the thermal energy resides is physically transported such as the movement of hot gas or hot liquid when it wants to rise because it is lower in density than the surrounding materials and experiences a buoyant force. This is natural convection. Convection can also be achieved with fans and blowers.

Practical heating in homes relies largely on convection to transport the high temperature gases throughout the room.

11.7 Radiation

The third method of transporting thermal energy involves radiation. A object with temperature T radiates energy in the form of electromagnetic waves. While all wavelengths are present in radiation, the intensity peaks at a particular wavelength which is characteristic for that temperature. This is known as Wien's law. The total radiation emitted is given by Stephan's law and depends not only on temperature, but surface area as well. The energy radiated per second or power is given by

$$P = \sigma A e T^4 \qquad (11.7)$$

where σ is a constant $= 5.6696 \times 10^{-8}$ W/m$^2 \cdot$ K^4, e is the emissivity and T is temperature in kelvin.

If energy is being absorbed because of the surroundings at temperature T_0, then the net radiation emitted is

$$P_{net} = \sigma A e (T^4 - T_0^4) \qquad (11.8)$$

11.8 Hindering Heat Transfer

Specially designed containers are needed to keep liquid helium and various gases that have been cooled until they liquefy. Dewar flasks and double Dewar flasks are used.

Vacuums between surfaces are used to control losses due to convection. Special coatings can help prevent radiation losses. Highly reflective surfaces will not absorb radiation, so they tend to stay cool inside.

11.9 Concept Statements and Questions

1. Internal energy of a system can exist in many forms including potential energy and kinetic energy. That

portion of internal energy which can cause temperature change when the amount is changed is called thermal energy.

2. Could power be given in units of calories/second? If so, estimate the power of your body by your time rate of Calorie usage. (How many Calories do you consume in a day? (Guess!)

3. Which takes more heat to change the temperature of equal amounts of mass of a material, high specific heat or low specific heat materials?

4. If you lift one kg a vertical height of 1 meter the work done is 9.8 J. About how many calories is this equal to? (Food calories are Calories or kilo-calories.)

5. Does it take more calories to melt a gram of ice or to vaporize a gram of water? Compare and see.

6. If you want a good insulting window would you prefer to buy one with an R rating of 5 or 15?

7. How does most of the energy from the sun reach the earth, by conduction, convection or radiation?

8. Do you get more heat from a campfire as you stand around it or from conduction or convection?

9. If two stars are identical except for temperature and star B is twice as hot as star A, how will power radiated by star B compare with star A? (16 times). The power radiated by a star is called its luminosity.

10. Give an example in which thermal energy transferred to a system does not increase the temperature of

the system, then give an example in which thermal energy of a system increases when there is no transfer of thermal energy to the system. Does the temperature change?

11.10 Hints for Solving the Problems

General Hints

1. Be careful not to mix unit systems. BTU units go with Fahrenheit, pounds etc. while calories or joules go with grams (or kilograms) and Celsius degrees. Follow the strategies for solving calorimetric problems given in section 14.3.

2. If P is given as a function of V, then PdV can be integrated or the area under the P,V curve can be obtained in some other way to get the work done.

3. Study the examples carefully to help prepare you for working the problems on your own.

4. Food energy is rated in Calories (1000 cal)

5. Refer to the problem solving strategy in section 11.4 for dealing with calorimetry problems.

Hints for Solving Selected Problems

1-12. Use the mechanical equivalent of heat to convert back and forth between mechanical energy added to a system and the equivalent heat. Also to calculate temperature change for a certain amount of heat added, use the equation that involves the specific heat and look up the specific heat for the substance given.

13-22. Use conservation of thermal energy transfer and the temperature change equation for heat added.

13-34. Again use conservation of energy, but this time heat may cause a phase change rather than a temperature change, so you have to use the latent heat formula for the phase change portion. Compute the energy involved in the phase change to see if there is any remaining to cause a temperature change.

35-42. Use the energy transfer equation for conduction for these problems including the R values given in tables to get the conduction rate.

43-47. These are power radiation and absorption problems requiring Stephan's law. Temperatures must be in kelvins.

12
The Laws of Thermodynamics

**Using Thermal
Energy to do
Work**

Understanding the laws of thermodynamics allows us to use thermal energy in a practical way. The first law is a statement about the conservation of energy and the second law is a statement about entropy which usually increases and can only decrease when energy is expended to an outside system. An understanding of this law lets us determine the practical limits of heat engines and refrigerators.

12.1 Heat and Internal Energy

Thermal energy is that portion of internal energy that changes with temperature. Also when two systems of different temperature are put into thermal contact, thermal energy will transfer as heat from the higher temperature to the lower temperature until both systems are at the same temperature. We then say equilibrium has been achieved.

It is important to realize that energy can be transferred between two systems even when no heat is involved. For example, two sticks rubbed together create additional thermal energy in each stick. Also a transformation of one form of energy to another can increase the thermal energy of the system. Nuclear and chemical forms of energy can be converted to thermal energy. Also frictional forces can put thermal inergy into the system. It is an energy transfer due to work.

12.2 Work and Heat

A macroscopic system that is internally in equilibrium can be described with the equation of state $PV = nRT$. These variables can be related to mechanical situations if we notice that work $W = F\Delta x = (F/A)A\Delta x = P\Delta V$, so

$$W = P\Delta V \qquad \text{(12. 1)}$$

If the pressure of a system remains constant, then the amount of work it does is just $P(V_f - V_i) = P\Delta V$. When P varies the work equals the area under the curve of a P versus V diagram. Also the work done depends upon the path between the initial and final states. (We have already discussed how friction and processes involving heat are not conservative as far as mechanical energy is concerned.) Both heat and work are processes for transfering energy so neither are conserved in thermodynamic processes.

When there is no heat transfer to the system, we call the processes involved *adiabatic processes.*

12.3 The First Law of Thermodynamics

If we denote the internal energy of the system with the letter U, the heat with Q and the work done by the system with W, then $\Delta U = Q - W$, which is a statement about the conservation of energy of the system and is known as *the first law of thermodynamics*. Thus we have

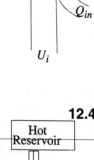

$$\Delta U = U_f - U_i = Q - W \qquad \text{(12.2)}$$

Note: It is the change in U that counts and we don't need to know the details of just how the internal energy changed. For isolated systems U remains constant. If there are cyclic processes so we come back to the initial state, even though the system is not isolated, the change in U is zero so $Q = W$.

12.4 Heat Engines and The Second Law of Thermodynamics

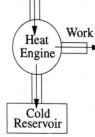

A heat engine is a device that converts thermal energy to other useful forms such as electrical or mechanical energy.

A heat engine is represented schematically in the figure. A substance goes through a cycle of absorbing and rejecting heat and doing work. The work is equal to the difference in the heat in and heat out, i.e. the net heat ,

$$W = Q_h - Q_c \qquad \text{(12.3)}$$

For a gas the work done is the area enclosed by the curve in a P,V diagram. The thermal efficiency is defined as the ratio of the net work done to the heat absorbed during one cycle, i.e.

$$e = \frac{W}{Q_h} = \frac{Q_h - Q_c}{Q_h} = 1 - \frac{Q_c}{Q_h} \qquad \text{(12.4)}$$

One way of stating the second law is: *It is impossible to construct a heat engine that, operating in a cycle, produces no other effect than the absorption of heat from a reservoir and the performance of an equal amount of work.*

12.5 Reversible and Irreversible Processes

Broken egg Irreversible process

A reversible process is one that leaves the system and the environment unchanged in going through a cycle. Otherwise the process is irreversible. All natural processes are known to be irreversible. If the changes occur very slowly so the environment and system are nearly always in equilibrium, the process is nearly reversible. There can be no dissipative effects present.

12.6 The Carnot Engine

Sadi Carnot in 1824 using the theory of thermodynamics described a cycle for a heat engine that has the highest efficiency possible. No other engine can get more work out of an input of heat than the Carnot engine. The cycle is bounded by four reversible processes, two of which are adiabatic (no heat lost or gained) and two are isothermal (no change in temperature). He showed the efficiency of the cycle to be

$$e_c = \frac{T_h - T_c}{T_h} = 1 - \frac{T_c}{T_h} \qquad \text{(12.5)}$$

In this equation T *must be in absolute temperature units, e.g. kelvins.* The higher the temperature of the input heat and the lower the temperature of the output heat, the higher the efficiency will be. Real engines have frictional and heat losses so they are never quite as efficient as a theoretical Carnot engine.

12.7 Heat Pumps and Refrigerators

A heat pump operates just oppositely to a refrigerator. The pump brings in heat from the outside ground or air and the deposits it in the enclosure. A refrigerator takes heat from the enclosure and deposits it in the surroundings. The coefficient of performance (COP) is given by

$$\text{COP(heat pump)} = \frac{\text{heat transferred}}{\text{work done by pump}} = \frac{Q_h}{W} \qquad \text{(12.6)}$$

A heat engine run through the Carnot-cycle has the highest possible COP which is given by $T_h/(T_h\text{-}T_c)$.

The COP for a refrigerator is Q_c/W and the highest possible COP is $T_c/(T_h\text{-}T_c)$.

12.8 Entropy

Temperature is a state function associated with the zeroth law of thermodynamics. Energy and its conservation is another state function and is associated with the first law. Another state function is entropy and it is associated with the second law. Entropy has to do with the amount of order or disorder of a system.

The Clausius definition of **change** in entropy is

$$\Delta S = \frac{\Delta Q_r}{T} \tag{12.7}$$

where the subscript r refers to reversible processes. One of the main outcomes of the concept of entropy is that real processes are not reversible and they increase entropy. The system tends to more disorder and entropy is a measure of that disorder. The change of entropy for a reversible cycle, such as a Carnot engine cycle, is zero. But all natural processes increase entropy and the entropy of the universe is constantly increasing.

12.9 Entropy and Disorder

Ordered States

Disordered States

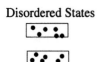

Entropy is a state function and, therefore, is dependent only upon the state of a system and not how the system got there. Thus, while most processes are irreversible, in calculating the change in entropy between two states we can devise a way of calculating over infinitesimal differences that are reversible so we can use the definition we have, namely $\Delta S = \Delta Q_r/T$. Thus the change in entropy can be determined from the change of state of the system regardless of how that change occurred.

Isolated systems tend toward disorder, and entropy is a measure of that disorder.

Disordered states in nature are much more probable than ordered states because there are many more of them, that is, the number of ways we can have disorder is much higher than the number of ways we can have order. Thus disorder is more probable, so entropy measures the degree to which a system has progressed towards the most probabilistic state.

In light of this view of entropy it is found that entropy can be related to the probability, W, of a system being in a particular state so that

$$S = k_B \, ln\mathrm{W}$$

(12. 8)

where k is Boltzmann's constant ($k_B = 1.38 \times 10^{-23}$ J/K) We are able to conclude from this that entropy is a measure of microscopic disorder.

The second law of thermodynamics is really a statement of what is most probable, not of what must be.

12.10 Concept Statements and Questions

1. Internal energy of a system can also include potential energy as well as kinetic energy of various forms, and such energy is considered to be thermal energy.

2. Are there some situations when the first law of thermodynamics can be broken? Give an example if it is possible.

3. How do pressure, P, and volume, V, relate to work done on a gas or by a gas?

4. What is meant by an "irreversible process"? How could it be made reversible? Why are all processes in nature irreversible according to the definition?

5. It is impossible to make a heat engine more efficient than the Carnot engine. What effect does temperature have on the efficiency of an engine? Which law or laws of thermodyanamics rules out perpetual motion machines?

6. Thermal energy will not naturally flow from a cold object to one with a higher temperature.

7. Is the coefficient of performance, COP, the same thing as efficiency? If not how does it differ?

8. Entropy always increases in the universe. If it decreases in a system it is because energy or work is put into that system which causes an even greater increase in entropy elsewhere.

12.11 Hints for Solving the Problems

General Hints

1. Be careful not to mix unit systems. BTU units go with Fahrenheit, pounds etc. while calories go with grams (or kilograms) and Celsius degrees. Follow the strategies for solving calorimetric problems given in section 14.3.

2. If P is given as a function of V then the area under the P,V curve can be obtained in some way to get the work done.

3. Study the examples carefully to help prepare you for working the problems on your own.

4. Food energy is rated in Calories (kilo-calories)

Hints for Solving Selected Problems

1-9. The area under a P,V diagram gives the work. Also you may need the relationship of $PV = nRT$ for an ideal gas. $P\Delta V$ gives work but $V\Delta P$ doesn't.

10-24. If work is done on a gas the internal energy of the gas increases. If the gas does work, the internal energy decreases. Study examples 12.2 and 12.3.

25-37. Use the equations developed for the efficiency of the ideal heat engine (Carnot engine). This sets the limit for performance of all other engines. You can work with either the heat transferred or the absolute temperatures to get the efficiencies. There is a formula for each.

38-43. Calculate the heat transferred for the two states of the system, and divide by T to get entropy change. In problem 42, the loss in KE could be considered the heat transferred.

44-46. Once you have figured the probability according to the problem, the entropy is obtainable from $S = k_B \ln W$.

13
Vibrations and Waves

**Periodic Motion
and Distortions
of the Medium**

Motion that repeats itself over and over is very common and applies to rotation, to oscillating springs, vibrating reeds, to waves and many others. Whenever there is a restoring force back to equilibrium for a displacement we get oscillatory motion of one type or another. The oscillations can be simple (without change from cycle to cycle), with diminished amplitudes (damping forces are present), or increased amplitudes (resonance phenomenon in forced oscillations), or combinations and variations of the above such as coupled oscillations, etc.

13.1 Hooke's Law

Many forces encountered in every day living are elastic in nature as long as the elastic limit has not been exceeded, for example, the slight bending of a rod or plank (like a diving board), the stretch of a spring or even a straight wire, the bounce in a tire full of air, and etc. For all these situations, the force causing the stretch or bend is proportional to the displacement. The reaction force is a restoring force trying to return everything back to equilibrium. The restoring force is then,

$$F_x = -kx \qquad \text{(13.1)}$$

which is known as Hooke's law. Simple harmonic motion (SHM) occurs with oscillating things that obey Hooke's law. The motion is simple and each cycle is like the one before. It should be noted, however, that not all repetitive motion (simple or not) is SHM.

In SHM the period, T, is the time to complete a cycle, the maximum displacement over a cycle is the amplitude, A, and the frequency, f, is the number of cycles/s. One cycle per second is called a hertz (symbol Hz).

Since $F = ma$, then the acceleration at any time is obtained from $ma = -kx$ and hence,

$$a = -\frac{k}{m}x \qquad \text{(13.2)}$$

13.2 Elastic Potential Energy

The potential energy, PE, in a system obeying Hooke's law depends on the square of the displacement with the initial value defined as zero when at equilibrium. By this definition of zero PE we define it as

$$PE_s = \frac{1}{2}kx^2 \qquad \text{(13.3)}$$

where k is the spring constant, i.e. the proportionality constant in Hooke's law, $F = -kx$.

This potential energy is conservative so we can include it as part of the mechanical energy in our conservation of energy equation, namely,

$$(KE + PE_g + PE_s)_i = (KE + PE_g + PE_s)_f \qquad \text{(13.4)}$$

If nonconservative, dissipative forces are present the work done by these forces is the energy lost from the total mechanical energy between the initial and final states, i.e. $W_{nc} = E_f - E_i$.

13.3 Velocity as a Function of Position

Using the conservation of energy equation from the previous section and concerning ourselves only with kinetic energy and elastic potential energy, we know the total energy at maximum displacement is $kA^2/2$ so

$$\frac{1}{2}kA^2 = \frac{1}{2}mv^2 + \frac{1}{2}kx^2 \quad \text{and solving for v}$$

$$v = \pm\sqrt{\frac{k}{m}(A^2 - x^2)} \qquad \text{(13.5)}$$

13.4 Comparing Simple Harmonic Motion with Circular Motion

The vertical up and down motion of a peg on the rim of a uniformly rotating wheel is SHM. The same is true for the horizontal back and forth motion. The two dimensional rotating motion when projected into a one dimensional plane is SHM.

PERIOD AND FREQUENCY

The period of the projection of rotation, T, is the same as the period of rotation, T. Thus the velocity on the rim of the wheel is $v_0 = 2\pi A/T$, which is the maximum velocity of the projection. Thus, $mv_0^2/2 = kA^2/2$ so

$$\frac{A}{v_0} = \sqrt{\frac{m}{k}} \quad \text{and so} \quad T = 2\pi\sqrt{\frac{m}{k}} \qquad \text{(13. 6)}$$

and the frequency, f, is simply $1/T$.

13.5 Position as a Function of Time

The position of a particle as a function of time for SHM is given by

$$x = A\cos(\omega t) \qquad \text{(13. 7)}$$

where $\omega = 2\pi f$ and the particle is at A when $t = 0$.

13.6 Motion of a Pendulum

The simple pendulum is an example of SHM when the amplitude of swing is small. The restoring force is really $F_s = -mg\sin\theta$, but when θ is small $\sin\theta \approx \theta$, so $F_g \approx -mg\theta = -mgs/L$, where s is the arc length and L is the length of the string. This equation is of the right form to be Hooke's law where the constant $k = mg/L$. By adapting the equation for period, T, from section 13.4, we get

$$T = 2\pi\sqrt{\frac{m}{(mg)/L}} = 2\pi\sqrt{\frac{L}{g}} \qquad \text{(13. 8)}$$

Thus we see that the period of the pendulum does not depend upon its mass, but only the length, L, of the pendulum.

13.7 Damped Oscillations

When resistive or frictional forces are present along with Hooke's law forces, and no external energy sources restore the lost energy due to the nonconservative forces, then the amplitude of each oscillation is diminished. We call this motion damped oscillations. The amount of damping depends upon how great the resistive forces are.

13.8 Wave Motion

Mechanical waves are disturbances in a medium which propagate through the medium. Examples are sound waves, earthquake waves, water waves etc. The particles in the medium may move up and down or back and forth or in some other complicated way as the wave passes, but they do not move with the wave. The source of these waves is some vibrating object. Other types of waves such as electromagnetic waves only need the vacuum in which to propagate.

[Curiously, the particles of the electromagnetic waves (photons = quantum of energy) do propagate with the waves. Also on the microscopic level we find still another type of wave, a probability wave associated with the motion of any and all particles.]

13.9 Types of Waves

When particles oscillate perpendicularly or transverse to the motion of the wave, it is said to be a *transverse wave*. The wave of a plucked string is transverse. If the wave oscillates parallel to the wave motion it is called transverse. Sound waves in air are longitudinal. The peak of the oscillation (maximum displacement from equilibrium) is called the *crest* and the minimum is called the *trough*. Water waves are neither exactly transverse nor longitudinal, but rather a combination of both as the particles move in an elliptical path.

13.10 Frequency, Amplitude and Wavelength

The distance from crest to crest or from trough to trough of a wave is called its wavelength. The wave propagates so that it travels a distance of one wavelength, λ, during one complete oscillation or cycle, which takes the period of time T. It is clear, therefore, that the speed of the wave is

$$v = \frac{\lambda \,(\text{wavelength})}{T \,(\text{period})} = \lambda f \qquad \text{(13.9)}$$

The height of the crest or maximum displacement is the amplitude of the wave.

13.11 The Speed of Waves on Strings

The speed of the wave on a string depends upon the mass per unit length, μ, of the string and the tension, F, in the string, thus

$$v = \sqrt{\frac{F}{\mu}}$$ (13. 10)

The speed depends upon elastic properties of the medium.

13.12 Superposition and Interference of waves

Linear waves obey the superposition principle which states:

If two or more traveling waves are moving through a medium, the resultant wave is found by adding together the displacements of the individual waves point by point.

Because of the superposition principle, two waves of the same type can pass through each other without being destroyed or even altered.

13.13 Reflection of Waves

Whenever a traveling wave pulse reaches a boundary, part or all of the pulse is reflected. Any part not reflected is said to be transmitted through the boundary or absorbed. If the boundary is fixed, the wave is entirely reflected but inverted (180° phase change). If the boundary is totally free or open, the wave is again 100% reflected but it is not inverted (0° phase change).

If the reflection goes to a boundary that is not totally fixed such as a heavier string, part of the wave will continue on but with a smaller amplitude and lower velocity, and part will be reflected and be inverted. If the string is lighter, the part going on will travel at a higher velocity and there will still be some reflected portion that is inverted.

13.14 Concept Statements and Questions

1. Simple harmonic motion (SMH) occurs for systems that obey Hooke's force law, $F = -kx$.

2. The potential energy in a spring or any other object obeying Hooke's law is conservative.

3. Simple pendulums and even physical pendulums approximate SHM when the amplitude is small.

4. Using conservation of mechanical energy one can determine the velocity of the object undergoing SHM at any position x.

5. Viewing a projection of a point on a uniformly rotating wheel shows it undergoes SHM. Hence SHM is simply a one-dimensional projection of circular motion. Does it matter onto which axis the projection occurs?

6. The inclusion of friction changes the SHM to damped oscillatory motion. The amount of damping can vary so the motion gradually decreases with decreasing amplitude and frequency, or drastically for strongly damped situations so the oscillations essentially disappear in one or two cycles.

7. The position of a particle versus time in SHM follows a cosine or sine function shape.

8. Traveling waves can be transverse or longitudinal. The medium in which the wave moves does not move with the wave but oscillates around an equilibrium point.

13.15 Hints for Solving the Problems

General Hints

1. For a simple pendulum the equation for SHM can be either in terms of the displacement angle, or the displacement distance, the distance along the arc.

2. Study the text examples.

3. Don't confuse frequency, f, and cycles per second with the angular frequency ω in radians per second.

Hints for Solving Selected Problems

1-4. You can practice using $F = -kx$ with these problems.

5-10. Use the formula for elastic potential energy and sometimes you need to use conservation of energy to relate velocity and position together.

11-19. Use the total energy formula ($kA^2/2$). Also you have a formula for v in terms of x for position other than equilibrium and end points. Look at example 13.5.

20-28. You may need to remember $v = r\omega = 2\pi f$.

24-28. k can be obtained from maximum amplitude and total energy. After getting k you can get the period or frequency from Equation 13.6.

34-39. The formula for the period of the pendulum will be useful. When comparing two pendulums, consider looking at the ratios of formulas. That way the constants drop out.

40-49. Use $\lambda f = v$ in all its variations. Once you have v you can get distance or time as needed.

50-56. Use the formula for the velocity of a wave on the spring. Also remember what happens to waves upon reflection. Review chapter material as needed.

14

Sound

**The Physics of
Sound Waves**

Sound waves are compressional waves by nature. The medium is usually thought of as air but can be any fluid or solid. In a solid the waves can also be transverse.

14.1 Producing a Sound Wave

Whatever the nature of a sound wave, its origin is always due to something vibrating. If the vibration occurs in a fluid such as air, the back and forth motion alternately produces compressions and rarefactions in the air along the direction of propagation of the wave. This process is superimposed upon the general random motion of the molecules in the medium.

14.2 Characteristics of Sound Waves

Compression and rarefaction are terms used in connection with sound waves. The compression corresponds to maximum pressure change (the crest) and rarefaction is also a maximum pressure change but is a trough. Being longitudinal the oscillations move back and forth in the direction of propagation.

CATEGORIES OF SOUND WAVES

The human ear can hear frequencies from 20-20 000 Hz (a typical good ear). Lower frequencies are felt but not heard and are called *infrasonic waves,* whereas frequencies higher than 20 000 Hz are the ones we designate as ultrasound.

Transverse waves occur when the disturbance is perpendicular to the direction of travel of the wave.

14.3 The Speed of Sound

The elasticity in the compressibility of the medium and the general inertial properties determines the speed with which the waves travel. To get the speed, we look at the bulk modulus, B, for fluids and Young's modulus, Y, for solids so that

$$\text{In Fluids} \quad v = \sqrt{\frac{B}{\rho}} \quad \text{and}$$

$$\text{In Solids} \quad v = \sqrt{\frac{Y}{\rho}}$$

(14.1)

The speed of sound also depends upon temperature. In degrees Celsius for air it varies according to

$$v = (331 \text{ m/s}) \sqrt{1 + \frac{T}{273}}$$

(14.2)

14.4 Energy and Intensity of Sound Waves

Intensity of a sound wave is defined to be the rate of flow of energy through a unit area perpendicular to the direction of the waves motion, thus

$$I = \frac{\Delta E}{\Delta t}\left(\frac{1}{A}\right) = \frac{Power}{Area}\left(\frac{W}{m^2}\right) \tag{14.3}$$

The range of intensity (as pressure amplitude) for the human ear is 3×10^{-5} Pa to 29 Pa over atmospheric, while the range of amplitude is 1×10^{-11} m to 1×10^{-5} m.

INTENSITY LEVELS IN DECIBELS

The loudest to the faintest sound signal that the human ear can hear varies over the enormous range of about 10^{12}. It is convenient, therefore, to use a logarithmic scale called decibels to compare intensities. Intensity in decibels is given by

$$\beta = 10 \log\left[\frac{I}{I_0}\right] \tag{14.4}$$

I_0 is the reference intensity taken to be 10^{-12} W/m2 at the threshold of hearing. Based on decibels, however, threshold of human hearing is 1 and the approximate maximum is 120 dB with which is associated pain in the ears.

14.5 Spherical and Plane Waves

For sound sources that propagate out in a spherical distribution, the intensity of the wave diminishes according to the inverse of the increasing surface area of the sphere, so

$$I = \frac{\text{average power}}{\text{area}} = \frac{P_{av}}{A} = \frac{P_{av}}{4\pi r^2} \qquad \text{(14.5)}$$

At large distances compared to the wavelength the spherical spreading of the waves becomes less and less noticeable and we approximate a *plane wave*.

14.6 The Doppler Effect

Relative motion alters the frequency and wavelength of a wave as it passes an observer. The alteration for mechanical waves depends also on whether the source or the observer is in motion relative to the medium. For light and other electromagnetic waves again the wavelength and frequency change if the observer and source are in relative motion, but since there is no medium for the wave to travel through, only the relative motion counts. This change in the waves due to motion is called the Doppler effect, named after Christian Johann Doppler, an Austrian physicist.

For an observer moving toward the source we get

$$f' = f\left(\frac{v + v_0}{v}\right) \qquad \text{(14.6)}$$

and for the source moving toward the observer

$$f' = f\left(\frac{v}{v - v_s}\right) \tag{14.7}$$

With both the source and observer in motion relative to the medium and allowing for motions of recession as well as approach, a formula that says it all is

$$f' = f\left(\frac{v \pm v_0}{v \mp v_s}\right) \tag{14.8}$$

In this formula the upper sign goes with velocity of approach and the lower sign goes with recession.

SHOCK WAVES

When the source is traveling faster than the velocity of the wave in the medium a wave-front in the shape of a cone is developed with the source at the apex. This cone is called a shock wave. The apex angle of the cone is given from the geometry as

$$\sin\theta = v/v_s \tag{14.9}$$

The bow wave of a boat is an example of such a wave in two dimensions and occurs when the boat moves faster through the water than water waves travel.

14.7 Interference in Sound Waves

Waves can interfere with each other and so also interference of sound waves is possible. When the crest of one wave superimposes upon the crest of another wave we get constructive interference, but if a crest superimposes onto a trough, then destructive inference occurs.

Splitting sound waves is also possible and each split-wave can be made to travel different paths and then brought back together in phase or out of phase and thus varying degrees of interference may be achieved.

14.8 Standing Waves

Strings with waves traveling in both directions and that are clamped on both ends totally reflect these waves. "Standing waves" can be set up when waves of equal amplitude are going in opposite directions but inverse of each other and reflecting off the ends with no amplitude variation. Where the waves cancel is called a node and the positions of maximum superimposed amplitudes are called anti-nodes. In a standing wave all points on the string oscillate vertically with the same frequency but the amplitude varies from zero to maximum.

Natural frequencies (normal modes) correspond to those frequencies that make it possible to have nodes at each end as well as in between for a string fixed at each end. This would be possible if half the wavelength is L (L = length of string) and again for

$$\lambda_n = 2L/n \quad \text{where} \quad n = 1, 2, 3, \dots \tag{14.10}$$

The corresponding frequencies are, therefore,

$$f_n = \frac{v}{\lambda_n} = \frac{n}{2L} v = \frac{n}{2L} \sqrt{\frac{F}{\mu}}, \qquad n = 1, 2, 3, \dots \tag{14.11}$$

The lowest frequency ($n = 1$) is called the fundamental or first harmonic, the next ($n = 2$) is called the second harmonic, and so on. Several different harmonics can be excited simultaneously and they all blend to give harmonious and usually pleasant sounds.

14.9 Forced Vibrations and Resonance

If a periodic force is applied to the stretched string and the frequency is near a natural frequency, the amplitude of motion of the system can become very large. This is called *resonance*. The natural frequencies of vibration of a system are often called resonant frequencies.

14.10 Standing Waves in Air Columns

Just as standing waves can be set up in a string, so can standing longitudinal waves (sound waves) be established in a column of air. If the column is closed, we get a displacement node (change in pressure antinode). If it is open, we get a displacement antinode (pressure change node)

The natural frequencies for a pipe open at both ends is

$$f_n = n\frac{v}{2L} \qquad n = 1, 2, 3, \ldots \qquad \text{(14. 12)}$$

If the pipe is closed at one end we get

$$f_n = n\frac{v}{4L} \qquad n = 1, 2, 3, \ldots \qquad \text{(14. 13)}$$

14.11 Beats

A beat is the periodic variation in amplitude of the sum of two superimposed waves of slightly different frequencies. The sum wave is heard as a sound with a frequency that is the average of the two component frequencies. The sum varies in intensity with a frequency that is the average of the component waves. The difference in frequency of the two waves is called the beat frequency.

14.12 Quality of Sound

The quality of sound of a musical instrument is also called timbre and depends on the number of harmonics present when a note is played and the relative intensity of each as compared to the fundamental.

14.13 The Ear

The human ear is divided into three regions: the outer and visible ear, the middle ear, and the inner ear. The outer ear collects the sound waves and concentrates them onto the eardrum which vibrates in phase with the sound waves. The middle ear transmits the vibrations on the eardrum with three small bones called the hammer, anvil and stirrup to the inner ear where they are sensed and transmitted to the brain for interpretation. The response of the ear varies with frequency and intensity. The ear has its own built-in protection against loud sounds providing they are not sudden.

14.14 Concept Statements and Questions

1. Can sound travel through water? How about steel? If it can, how would the speeds of travel compare with the speed of sound in air?

2. The things that characterize a wave are its wavelength, velocity, frequency, shape and whether it is longitudinal or transverse. How are the first three related?

3. The intensity of a wave is the power per unit area and depends on the square of the amplitude and the square of the frequency.

4. The intensity of a spherical wave diminishes as $1/r^2$. The intensity of plane waves does not diminish with distance (providing no dissipative elements are present).

5. The decibel is a unit of sound level, which is a logarithmic measure of relative intensity.

6. Relative motion with respect to the medium of the source or observer gives a doppler change in wavelength and frequency.

7. When the source of a sound wave is traveling faster than the wave (supersonic) a shock wave is produced. At what angle does the shock wave travel?

8. To produce a standing wave it is necessary to have two identical waves traveling in opposite directions.

9. Transverse waves in a string fixed at both ends vibrate at natural frequencies with nodes at the end

points. These normal modes can be considered as traveling waves and their reflections.

10. Waves of slightly different frequencies produce beats or modulations in the intensity and the frequency of the beats is simply the difference in the two frequencies.

14.15 Hints for Solving the Problems

General Hints

1. To obtain the speed of a sound wave use the temperature dependent formula 14.2.

2. Review logarithms. Remember $\log 10 = 1$, $\log 100 = 2$, $\log 1000 = 3$, etc. Also recall that $\log r^2 = 2 \log r$.

3. Study the examples.

4. For the Doppler effect you need to carefully study the signs associated with approaching as opposed to departing signals and motions of the oberver and source. This is the hard part about such problems.

Hints for Solving Selected Problems

1-8. Use equation 14.2 to get the speed of sound at a particular temperature. Remember T is in Celsius and not kelvins.

9-10, 12. Work with the decibel formula. .

11, 14. Power equals intensity times area. Intensity comes from decibel equation. The ratio of intensities equals the ratio of powers when area is constant.

15-19. There is a large variety of problems here. Several involve computing the decrease in intensity with distance. It decreases according to $1/r^2$.

20-30, These problems include the Doppler effect in addition to variations in intensity with respect to distance.

31-36. Destructive interference results when the path difference is one-half wavelength and constructive interference occurs when the path lengths differ by an integral number of wavelengths. The wavelength can be obtained from the frequency and speed of the wave.

37-44. For standing waves the wavelength of the wave can be obtained from the geometry of the system. See the formula that relates wavelength to length of string. Remember the velocity is related to the tension in the string by $v = \sqrt{\dfrac{F}{\mu}}$.

47-54. Apply the formulas that relate wavelength to air column length. The temperature dependence of the speed of sound needs to be taken into account for some problems as you did in problems 1-8.

55-57. The frequency of the beats is the difference between the individual frequencies. The frequency of the sum is the average, namely $2f_{av} = f_1 + f_2$.

15
Electric Forces and Electric Fields

The Force We Encounter the Most Frequently

Here we begin a study of the electric forces and the electric field to gain an understanding of them. In daily life we find the electrical force operating everywhere, because it is the force that holds most things together. Gravity is also common, but it is only capable of holding objects together where at least one is very massive, whereas the electrical force functions on the atomic level and is responsible for most of chemistry.

15.1 Properties of Electric Charges

Electric charges have the following properties.

1. Two kinds of charges exist in nature called positive and negative with the property that unlike charges attract one another and like charges repel one another.

2. The force between charges varies as the inverse square of their separation.

It is also known that (1) charge is conserved in all reactions and (2) charge is quantized with the unit of free charge being the charge on a proton or electron, $q_e = e$ = 1.602 x 10^{-19} coulomb = 1 electronic charge.

15.2 Insulators and Conductors

Conductors are materials in which electric charges move freely, and insulators are materials in which electric charges do not move freely.

Examples of good conductors includes all the various metals. Insulators are things like glass, wood, plastic, etc. Semiconductors have properties between those of conductors and insulators.

CHARGING BY INDUCTION

When a charged insulator is brought near a conductor the charges on the conductor redistribute themselves. Only the negative charges (the electrons) can move freely and they move so that unlike charges are as near as possible to the charge on the insulator and the like charges are as far away as possible. If a connection is made with the ground the negative charges rush into the ground or come out of the ground, so that if now the ground connection is broken, the conductor is left with a net charge. Removing the insulator and its charge leaves the conductor charged oppositely to the insulator. This is called charging by induction.

Charging by induction requires no contact with the object inducing the charge, but does require that a conducting path be made to ground and then broken.

15.3 Coulomb's Law

The electric force between two charges is given by Coulomb's law

$$F_{21} = \frac{k\lfloor q_1 \rfloor \lfloor q_2 \rfloor}{r^2}$$

(15. 1)

where k is called the Coulomb constant and $k = 8.9875 \times 10^9$ N·m²/C². If both charges have the same sign the force is repulsive. If the charges have opposite signs, the force is attractive.

Like gravity, the electric force can act through the vacuum of space. It is a field-force and can make its presence felt without contact of one object on another.

THE PRICIPLE OF SUPERPOSITION

The field due to a group of point charges is the sum of the vector fields due to each. Likewise the electric force on a charge is the sum of the vector forces due to each of the other charges present.

15.4 Experimental Verification of Coulomb's Force Law

Using a thin fiber with a thin rod attached and charged spheres on the ends of each rod, Coulomb was able to discover the $1/r^2$ nature of the electric force. Modern experiments of a different design have established it to a high accuracy.

15.5 Electric Fields

The magnitude of the electric field at the position of q_0 due to the charge q is the electric force F divided by q_0, or

$$E = \frac{F}{q_0} = \frac{k|q|}{r^2}$$ (15.2)

Because of the superposition principle, the field due to a group of point charges is the sum of the vector fields due to each.

15.6 Electric Field Lines

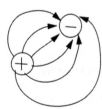

It is sometimes convenient to draw lines to help visualize the electric field. This is done by requiring

1. The electric field vector, **E**, is tangent to the electric field line at each point.

2. The number of lines per unit area through a surface that is perpendicular to the lines is equal to the strength of the electric field in that region. The field is greatest where the density of lines is greatest.

3. The lines must begin on positive charges and terminate on negative charges or, in the case of an excess of charge, begin or terminate at infinity.

4. The number of lines originating or terminating on a charge is proportional to the amount of charge, and

5. No two field lines can cross or touch each other.

Electric flux is a measure of the number of electric field lines penetrating some surface, i.e. the density of field lines per unit area is an indicator of the electric flux. The number of lines coming from or terminating on a charge is independent of the shape of the surface enclosing the charge

15.7 Conductors in Electrostatic Equilibrium

Charges are free to move in a conductor. When all the charges have come to rest we have the electrostatic equilibrium situation. For an isolated conductor we note the following characteristics:

1. The electric field is zero everywhere inside the conductor. (If it weren't, there would be forces that would cause the charges to move and we wouldn't be able to say that we were at equilibrium)

2. Any excess charge resides entirely on the outer surface.

3. The electric field just beyond the outside surface is perpendicular to the surface.

4. The charge tends to accumulate at locations of small radius of curvature, i.e. sharp points.

15.8 The Millikan Oil-Drop Experiment

By supporting small droplets of oil with weight, mg, and carrying an electric charge, $-q$, with an electric field, E, Robert Millikan was able to determine the magnitude of the smallest unit of charge. He thus measured the magnitude of the charge on a single electron or single proton and determined it to be 1.60×10^{-19} C.

15.9 The Van de Graff Generator

-Robert J. Van de Graff conceived of and built a genera-tor that takes advantage of the absence of electric field in a region enclosed by a conductor to build up a large charge on the outside of the conductor. He transported charge on a moving belt to the inside of a conducting dome or sphere where the charges were easily transferred from the belt to the dome since $E = 0$ inside, but the charge would then move to the outside of the dome creating a high E-field and electric potential there.

15.10 The Oscilloscope

The oscilloscope uses a tube called the cathode ray tube, CRT, which has a beam of electrons (cathode rays.) The beam can then be deflected with electric and magnetic fields. The beam hits a phosphor-surface which glows wherever hit by the electron beam. Thus, input voltages and their variations are easily seen as deflections on the CRT screen.

15.11 Concept Statements and Questions

1. The Coulomb force is similar to the gravitational force, and so the definitions of their fields are also similar.

2. One must add the electric fields from several different charges in a vectorial sense to get a resultant field.

3. The number of electric field lines to be drawn in the representation of the electric field is proportional to the strength of the electric field in that region.

4. The number of field lines per unit area perpendicular to the lines is called the electric flux. The flux near a charge is proportional to the amount of the charge.

5. High electric potentials can be created using what kind of device?

15.12 Hints for Solving the Problems

General Hints

1. For problems involving Coulomb's law, make a sketch and mentally think out an approximate solution before you do any calculating. Thus you can anticipate approximately what your answer will be.

2. Draw in electric field lines going from positive to negative charge and get a visual clue for what is going on.

3. Working with vector sums can be difficult and is usually most easily handled by breaking the vectors into components and summing the components.

4. Study the problem-solving strategies in section 15.10 and all of the examples. They are very helpful.

Hints for Solving Selected Problems

1-44. Use Coulomb's law. Remember the electric force is a vector directed along the line connecting two charges. When there are more than two charges in a problem, calculate the force of electric field for each pair and then add them vectorially. The vector addition can best be accomplished by computing the components of the force and then adding components and finally getting a resultant.

45-52. In making sketches of electric field lines remember that they are proportional to the amount of charge, they begin on positive charges and end on negative charges. If there is an unbalance of charge, the extra lines end or begin at infinity.

53-56. The electric field inside a conductor is zero in the static situation.

16
Electrical Energy
and Capacitance

Electricity as a
Conservative
Force

The electrostatic force is a conservative force and, therefore, it is meaningful to talk about an electric potential energy. This in turn enables us to define an electrostatic potential, a concept simpler than force to work with in describing circuits and many other electrical phenomena.

Capacitors are devices used to store electric charge and usually involve conducting plates separated by an insulator.

16.1 Potential Difference and Electric Potential

Like gravity, the electrical force depends upon the inverse square of the distance between two charges. Thus like gravity, the force is conservative. The work done in moving a small test charge q in a uniform electric field E is simply $W = Fd = qEd$. And by definition, the work done on the charge by a conservative force equals the negative of the change in potential energy, ΔPE, so

$$\Delta PE = -W = -qEd \qquad \text{(16. 1)}$$

We define the potential as $V \equiv \dfrac{PE}{q_0}$ and, therefore,

$$\Delta V \equiv V_B - V_A = \frac{\Delta PE}{q} \qquad \text{(16. 2)}$$

The unit of electric potential is the volt where 1 volt = joule/Coulomb or J/C and electric field which is 1 N/C can now be expressed as volt/meter or V/m.

In a uniform field the difference in potential is simply $\Delta V = -Ed$. From this we can also get the change in electric potential energy, $\Delta PE = q_0 \Delta V = -q_0 Ed$. This equation applies when d is in the direction of $\mathbf{E}$. When d is 90^0 with respect to $\mathbf{E}$ there is no change in potential nor potential energy. Thus we have "surfaces", a whole family of them, of equal potential known as *"equipotential surfaces"*. **No work** is done in moving a charge along an equipotential surface.

16.2 Electric Potential and Electric Potential Energy Due to Point Charges

By inserting the equation for the electric field due to a point source into the equation for potential difference we obtain

$$V = \frac{kq}{r}$$

(16. 3)

According to this convention the electric potential for a point charge at infinite distance r, is zero. For a group of charges we get the scalar sum over i of the potential due to each charge q_i. If another charge is placed at a distance r from the first charge the potential energy is

$$PE = q_2 V_1 = k\frac{q_1 q_2}{r}$$

(16. 4)

For a group of charges the total energy in the system is the sum of the energy between each pair.

16.3 Potentials and Charged Conductors

Since the work required by the electric force to move a charge in a potential is $W = -\Delta PE$ and the change in potential energy is $\Delta PE = (V_B - V_A)$, by putting them together we get

$$W = -q(V_B - V_A)$$

(16. 5)

Thus if V_A and V_B are equal, no work is required in moving a charge from point A to point B. Since every point within a conductor when at electrostatic equilibrium, including the surface, is at the same potential throughout, no work is required to move any charges. This is also true whether charged or not. This also implies that the electric field is zero throughout. If it were not so, the charges would move to make it so, but we are considering electrostatic equilibrium, therefore, all charges are at rest.

The potential within a cavity of a conductor of arbitrary shape is constant, providing there are no charges within the cavity, and the electric field, therefore, is zero throughout the cavity. The conductor is an equipotential surface.

THE ELECTRON VOLT

A convenient unit of energy when working with charged particles in a potential is the electron volt. It is defined as the energy that an electronic charge of 1 gains when accelerated through a potential difference of one volt. It can be converted to joules by

$$1 \text{ eV} = 1.60 \times 10^{-19} \text{ J or } (C \cdot V) \tag{16.6}$$

16.4 Equipotential Surfaces

As the name implies, an equipotential surface is one which has the same potential at every point on the surface. Because the potential is a constant, it requires no work to move a charge from one point to another. Equipotential surfaces are always perpendicular to the electric field lines in the vicinity.

16.5 The Definition of Capacitance

Two conductors having a potential difference of V between them with equal and opposite charges, $+Q$ and $-Q$, constitute what is called a capacitor, and the capacitance of the capacitor is defined as

$$C \equiv \frac{Q}{V}$$

(16. 7)

Capacitors are storehouses for charge and energy. Capacitance is **always positive** and depends only upon the geometry of the capacitor. The units are named Farads where 1 Farad = 1 F = 1 C/V.

16.6 The Parallel-Plate Capacitor

For a parallel plate capacitor the voltage V is given by Ed where $E = Q/(\varepsilon_0 A)$. The capacitance, therefore, is

$$C = \frac{Q}{V} = \frac{\varepsilon_0 A}{d}$$

(16. 8)

A is the area of one of the plates, and ε_0 is a constant called the permittivity of free space, with the value $\varepsilon_0 = 8.85 \times 10^{-12}$ C^2/N·m^2 = $1/4\pi k$, where k is the Coulomb constant.

16.7 Combinations of Capacitors

In circuits capacitors can be found in series (one after the other), in parallel (having each of their sides connected to the other so as to be at equipotential), and in various combinations of these arrangements. We sometimes need to find the equivalent capacitor of the various combinations.

PARALLEL COMBINATION

For capacitors in parallel the total charge stored is the sum of the charges on each, namely $Q = Q_1 + Q_2$. But since $Q = CV$ for a capacitor we see $CV = C_1V_1 + C_2V_2$. For conductors in parallel it is also true that $V_1 = V_2 = V$, so the voltages cancel out and we get

$$C_{equiv} = C_1 + C_2 + C_3 + ...$$

(16. 9)

It is clear that the combination of capacitors in parallel gives an equivalent capacitance that is greater than any of the individual capacitances.

SERIES COMBINATION

In series (one after the other) each of the capacitors carries the same charge and the voltage is the sum of the voltages on each, so that $V = V_1 + V_2$. Substituting for the V's from $V = Q/C$ and cancelling the Q's gives us

$$\frac{1}{C_{equiv}} = \frac{1}{C_1} + \frac{1}{C_2} + \frac{1}{C_3} + ...$$

(16. 10)

This shows that the equivalent capacitance of a series combination of capacitors is always less than any individual capacitance in the combination.

16.8 Energy Stored in a Charged Capacitor

The energy in a charged capacitor is equal to the work required to charge it. The work required to transfer a small amount of charge ΔQ from the negative to the positive plate at potential V is $\Delta W = V\Delta Q$. Since V increases linearly as more charge is added, the total work must be the average V, which is $(1/2)(0 + V_f)$, times Q. So the total work done and stored in the capacitor as energy E is

$$W = E = \frac{QV}{2} = \frac{Q^2}{2C} = \frac{CV^2}{2}$$

(16. 11)

16.9 Capacitors with Dielectrics

A dielectric is a nonconductor. When placed between the plates of a capacitor, it increases the capacitance and the voltage decreases by a factor κ, so $V = V_0/\kappa$ and $C = \kappa C_0$. The capacitance of a parallel plate capacitor with a dielectric inserted is

$$C = \kappa \frac{\varepsilon_0 A}{d}$$

(16. 12)

κ is called the dielectric constant and is greater than 1. A dielectric in a capacitor has the following advantages:

1. It increases the capacitance of a capacitor.

2. It increases the maximum operating voltage of a capacitor

3. It may provide mechanical support between the conducting plates.

There are various types of commercial capacitors which include (1) paraffin-impregnated paper rolled up between thin conducting sheets into a cylinder, (2) interwoven metal plates immersed in silicone oil, (3) construction out of ceramic materials, (4) variable capacitors made of two interwoven sets of metal plates with one being moveable and (5) electrolytic capacitors (with thin metallic oxides on the surface serving as the dielectric).

THE STUD FINDER

The stud finder which locates the boards behind the wall covering utilizes a capacitor whose capacitance changes when the stud finder is near the board.

16.10 Concept Statements and Questions

1. Because the electric force is conservative, unique values of the potential energy difference between two points for electric fields can be given as well as the potential difference.

2. Electric potential energy (difference) can be found from the product of charge and electric potential (potential difference).

3. Surfaces perpendicular to the electric field have equal potentials and are called equipotential surfaces.

4. Potentials and potential energies are scalar quantities. As such, they can be used to solve some problems that would otherwise require the use of electric forces and electric fields, which are vectors. Working with scalars is usually much easier.

5. Conductors are equipotential regions throughout in an electrostatic situation. How much work is required to move a charge from one position to another in a conductor?

6. Capacitance depends only on the geometry of the capacitor.

7. What capacitor combination do you use if you want to increase the capacitance? if you want to decrease the capacitance?

8. Dielectric material inserted into a capacitor increases the capacitance and lowers the voltage.

16.11 Hints for Solving the Problems

General Hints

1. Study the hints and problem solving strategies given in sections 16.2 and 16.7 of the text. They are very helpful. Also study over the examples.

2. Capacitors in parallel are equivalent to one large capacitor, i.e. it is as though all the plates on one side are touching and similarly for the other side.

Hints for Solving Selected Problems

1-14. Using conservation of energy and the equivalence of work and energy, you can use the relationships developed among work, potential energy, potential and electric field to solve these problems.

15-27. The same comment applies here as in problems 1-14 above but here the electric field is due to a point charge. You can get the force from the field and then the acceleration from the force. Through the acceleration you can get velocity, position, and time. Also the kinetic energy of a particle can be obtained if you know the potential energy, qV, it loses or gains. The potentials, being scalars, add directly.

28-36. Use the definition of capacitance. Several of the problems are concerned with parallel plate capacitors, for which you have an equation in terms of its geometry. Once you cet C and then V, it is easy to get the electric field as V/d.

37-51. The addition of capacitance for both the parallel and series case is very straight forward. When you have a mixture, add various combinations until you have an equivalent series or equivalent parallel set of capacitors which you can also add together. (See examples in the text.)

52-56. Use the formulas for the energy stored in a capacitor and how it relates to charge, voltage, and capacitance.

57-62. Use the formulas involving the dielectric constant κ.

17
Current and Resistance

**Electric Charges
Moving in a
Conductor**

Until now we have considered only electrostatics. We now want to consider electrical current which is the rate of flow of charges in motion. Also of interest is the resistance to the free flow of charges in a conductor and what effect that resistance has.

17.1 Electric Current

If charges are flowing perpendicularly to a surface, the current is the amount of charge passing through that surface per unit time. We have both average current to consider and instantaneous current given as

$$I_{av} = \frac{\Delta Q}{\Delta t}, \quad \text{and} \quad I \equiv \frac{dQ}{dt} \qquad (17.1)$$

The unit of electric current is the ampere where 1 A = 1 C/s. Current is positive in the direction the positive charge moves or in the opposite direction to negative charge motion. Since electrons are negatively charged, positive current is opposite in direction to the flow of electrons.

17.2 Current and Drift Speed

The amount of charge in motion is the product of the number density of charge carriers, n, times the volume flowing past a surface per unit time, i.e., Av, times the amount of charge on each carrier, thus $\Delta Q = (nAv_d\Delta t)q$, where v_d is the drift velocity ($\sim 10^{-4}$ m/s) of the charge carriers. The current then becomes

$$I = \frac{\Delta Q}{\Delta t} = nqv_dA \qquad \text{(17.2)}$$

17.3 Resistance and Ohm's Law

In most conductors the current is proportional to the applied voltage, $V = IR$. R, the proportionality constant, is called the resistance of the conductor and the relationship is called Ohm's law. R is measured in units called ohms (Ω) and has the SI equivalent units of volts/amp.

Ohm's law is an empirical relationship and is not a fundamental law of nature. It is valid only for certain materials called *ohmic* materials and applies only over limited values for current and voltage. Nevertheless, the law is very useful and practical for many electrical circuits.

17.4 Resistivity

The resistance of an ohmic conducting wire is found to be proportional to its length and inversely proportional to the cross-sectional area A. The proportionality constant, ρ, is called the resistivity and its inverse σ is called the conductivity, and they depend upon the nature of the material in the conductor, thus

$$R = \rho \frac{l}{A} = \frac{l}{\sigma A} \qquad (17.3)$$

17.5 Temperature Variation of Resistance

The resistivity, ρ, in most metals increases linearly with increasing temperature over a limited temperature range according to

$$\rho = \rho_0 [1 + \alpha (T - T_0)] \qquad (17.4)$$

where α is the temperature coefficient of resistivity. Of course the resistance will experience a similar relationship so that

$$R = R_0 [1 + \alpha (T - T_0)] \qquad (17.5)$$

Most materials also have non-linear regions and semiconductors have negative temperature coefficients of resistivity.

17.6 Superconductors

Superconductors belong to a class of metals and compounds whose resistivity goes to zero below some critical temperature, T_c. Thousands of superconducting materials are known, but surprisingly some of our best conductors at room temperature such as copper, silver, and gold are not among them.

In superconductors a current persists without any applied voltage. The material with the highest critical temperature known so far with $T_c \approx 150$ K is a material containing an oxide of mercury. It is hoped that materials will be found with even higher values of T_c, can be drawn into wires, and hopefully also have other desirable properties.

17.7 Electrical Energy and Power

Since the energy change corresponding to a change in potential is $\Delta U = q\Delta V$, then the rate at which energy is lost as charge, Q, passes through a resistor is $\Delta U/\Delta t = (\Delta Q/\Delta t)\Delta V = IV$. Therefore, power loss in a circuit is

$$P = IV = I^2 R = \frac{V^2}{R} \qquad (17.6)$$

17.8 Energy Conversion in Household Circuits

Electrical power can be used to do work, or often times it is used to provide heat such as in electric-cooking ranges, heating of the home, hair dryers, etc. Work is done by vacuum cleaners pulling in dirty air, and by other household devices involving motion like grinders, blenders, garbage compactors, and dish washers.

The cost of electric power pennies per kilowatt-hour (kWh). A kWh is the power of 1000 watts used continuously for one hour. Thus

$$1 \text{ kWh} = (1000 \text{ W})(3600 \text{ s}) = 3.6 \times 10^6 \text{ J} \qquad (17.7)$$

17.9 Concept Statements and Questions

1. The rate of flow of charges is current. What is the convention used for the sign of the current?

2. Does Ohm's law always hold for currents and voltages in a circuit?

3. For materials which obey Ohm's law, the amount of resistance in a wire increases proportional to its length. The voltage required to maintain the same current, therefore, is also proportional to the length.

4. Resistance usually increases with increasing temperature, but this is not true for all materials.

5. Why are superconductors considered to be so wonderful? What good are they?

6. Look at the electrical power equation and determine what happens to the cost of electric power when current is doubled through a constant resistance.

17.10 Hints for Solving the Problems

General Hints

1. From table 17.1 you can get the resistivity of many different materials and from the resistivity you can figure the resistance. (See Example 17.4).

2. Longer lengths of a wire increases its resistance. Adding more length is like adding another resistor in series. The cross sectional area A of a wire decreases the resistance as it enlarges the path for the electrons to flow. Because it adds more paths for the current, it is similar to adding resistance in parallel.

Hints for Solving Selected Problems

1-8. Working with the basic definition of current and the formulas developed will suffice for working these problems.

9-28. Use Table 17.1 to get resistivity from which you can get the resistance if you know the length and cross-sectional area of the wire.

29--38. The temperature coefficients of resistivity are found in Table 17.1 and are needed to work these problems. Look at Example 17.5. Once you have the temperature variation of the resistivity, you can figure the change in resistance of the wire. The relationship between V, I and R in Ohm's law, plus the connection between resistance and resistivity, plus the dependence of resistance upon temperature will enable you to work these.

39-50. Remember $P = W/t = E/t$ in mechanics and $P = IV = I^2R$ in electrical circuits. This will help you to work these problems. Energy $= Pt = VIt$ etc.

18
Direct Current Circuits

Charge in Motion in one Direction

Direct currents are those which have the charge flowing in only one direction. The source that maintains a constant current, i.e., the "pump" that makes the charge move is a potential source called "emf". It is often a battery but could be a generator. We will work mostly with circuits of constant current.

18.1 Electromotive Force

The energy source used to maintain constant current in a circuit is called electromotive force (abbreviated "emf"). Sources of emf are devices like batteries, and generators that increase the potential energy of charges circulating in circuits.

Most batteries have internal resistance so the voltage V they deliver to a circuit will be the emf, ξ, across their terminals less the Ir loss due to the internal resistance.

$$V = \xi - Ir \tag{18.1}$$

Since we have $V = IR$, where R is the load resistance, the current can be solved in terms of the two resistances R and r. We get

$$I = \frac{\xi}{R + r} \tag{18.2}$$

The power dissipated by the source of the emf into the load and internal resistance is

$$I\xi = I^2 R + I^2 r \tag{18.3}$$

18.2 Resistors in Series

For resistors in series we have $V = IR_1 + IR_2 + \ldots = IR$, thus we see that resistors in series simply add, or

$$R = R_1 + R_2 + R_3 + \ldots \tag{18.4}$$

The equivalent resistance is greater than any individual resistor.

18.3 Resistors in Parallel

For resistors in parallel, the voltages across each resistor is the same but the currents add so $I = I_1 + I_2 + I_3 + \ldots$, thus

$$\frac{1}{R_{eqiv}} = \frac{1}{R_1} + \frac{1}{R_2} + \frac{1}{R_3} + \ldots \tag{18.5}$$

Therefore, the equivalent resistance of the combined resistances is less than the smallest resistance in the combination. The current has many paths through which to flow, so the resistance is reduced.

18.4 Kirchhoff's Rules and Simple DC Circuits

The procedure for analyzing complex circuits is greatly simplified by using some simple rules given to us by Kirchhoff. They are:

1. Junction rule: The sum of the currents entering any junction must equal the sum of the currents leaving that junction. (Equivalent to conservation of charge.)

2. Loop rule: The sum of the potential differences across each element around any closed circuit loop must be zero. (Equivalent to conservation of energy.)

The following two points are an aid in applying the loop rule. (1) If a resistor is traversed in the direction of the current, the change in potential across the resistor is $-IR$, whereas if it is traversed oppositely, the change is positive. (2) If a source of emf is traversed in the direction of the emf (from - to + on the terminal), the change in potential is + emf, otherwise, it is negative.

18.5 RC Circuits

If we include capacitors in our circuit, the current varies with time.

Charging a Capacitor

Following the loop rule of Kirchhoff for a simple circuit containing an emf, a resistor, and a capacitor, we get

$$\xi - \frac{q}{C} - IR = 0 \qquad \text{(18.6)}$$

A solution to this equation is

$$q(t) = C\xi(1 - e^{-t/RC}) = Q(1 - e^{-t/RC}) \qquad \text{(18.7)}$$

where RC has units of time and is called the time constant, $\tau = RC$.

When a capacitor is discarded, $\xi = 0$ and I is of opposite sign, so $IR = q/C$. The solution of this equation for charge as a function of time is

$$q(t) = Qe^{-t/RC} \qquad \text{(18.8)}$$

The voltage V across the capacitor decreases in a similar fashion so that $V = \xi e^{-t/RC}$, where ξ is the initial voltage.

18.6 Household Circuits

Electric motors, lights, and other devices are connected in parallel to the incoming power lines so that each system will operate at the same voltage (about 120 V). Each system uses the current needed to operate according to design. To protect the household wiring from drawing too much current and, therefore, over heating and burning out, circuit breakers or fuses are inserted into the lines. Some very heavy-duty appliances such as ovens, kitchen ranges, and electric dryers require

240 V instead of 120 V. The power company can provide this voltage difference by one line being 120 V above ground and the other 120 V below ground. Thus the difference is 240 V. The 240 V appliance only requires half the current as that of a 120 V system. Therefore, smaller diameter line wires are needed.

18.7 Electrical Safety

People and animals need to be protected from the possibility of electrical shock by insulation about the wires. If the insulation becomes worn and badly frayed, the power cord should be replaced. Electrical shock can occur whenever a person or animal comes in contact with a "live" wire and ground. Electrical shock can cause serious injury including severe burns and/or interruption of the proper functioning of vital organs such as the heart. These could lead to death of the organism. The seriousness of electrical shock depends upon the amount of current and the duration. Ground-fault interrupters (GFI) in household circuits sense small leakages of current to ground by comparing currents to and from a device. If these currents differ, the circuit is shut off.

18.8 Concept Statements and Questions

1. Direct current is produced by a source of electrical power that allows the current to flow in only one direction. When does the voltage output of a battery depend upon its internal resistance?

2. Given several resistors to use in a circuit, how would you arrange them to get the greatest possible resistance? The least?

3. Does Ohm's law always hold for currents and voltages in parallel and in series if it holds in each element separately?

4. An emf will increase the potential in a circuit. Which is greater, the emf or voltage V across the battery terminals? Can they ever be equal? When?

5. Kirchhoff's rules apply to all junctions and loops in a circuit.

6. Capacitors in a circuit cause the current to be a function of time. How does the current depend on time?

7. Resistances in parallel give a lower resistance, so why are houses wired with circuits in parallel?

8. What are the dangers of electrical shock?

18.9 Hints for Solving the Problems

General Hints

1. Look at the hints given in sections 18.3 and 18.4. Also the examples are very helpful.

2. Memorize Kirchhoff's rules. They're easy to learn because they make sense and are needed in solving circuit problems of all types.

Hints for Solving Selected Problems

1-16. Add resistors in series as $R_1 + R_2 + R_3 + \ldots$ and add the reciprocals for resistors in parallel. The current through resistors in series is the same in each resistor, but in parallel they add together to give the total. Look at the figures given to help understand what is going on and what is wanted.

17-30. Study the circuit diagrams in the figures very carefully. Apply Kirchhoff's rules for each circuit. Look at the sketches of the circuits and write down the corresponding equations. Then solve the equations simultaneously.

31-36. Use the equations for RC circuits. The time constant is simply $\tau = RC$.

37-42. You know the power is given by $P = IV$ and you know how to deal with both parallel and series circuits utilizing resistor. Use these equations as needed in each of these problems.

19
Magnetism

Another Aspect of Electricity

Magnetism is known to be connected with moving charges and is, therefore, closely connected to electricity. There are many useful devices using this phenomenon. In a way it is harnessing electricity in a package, the magnet.

19.1 Magnets

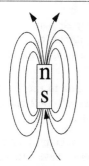

The discovery and use of magnetism goes back to about 800 B.C. Its connection with electricity was discovered in 1819 by Hans Oersted. James Clerk Maxwell was able to show that electricity and magnetism are two aspects of the same thing in his unified theory of electromagnetism.

Similar to electric charges, like magnetic poles repel and unlike poles attract. A major difference, however, is that magnetic poles have no real existence. They cannot be isolated like charges can nor can their precise position be defined.

Magnets can be made by putting materials that are susceptible to being magnetized, such as iron, in a magnetic field. Stroking the iron on a strong magnet, thumping it, or heating it can increase the amount of magnetism the iron piece takes on.

Besides magnetic fields associated with certain materials, a charged particle in motion also has a magnetic field.

The direction of any magnetic field is defined as the direction a north pole would move if placed in the field.

The practical applications of using magnets in industry are very numerous from electromagnets to uses in tape recorders, cassette players, magnetic tapes, floppy disks, etc.

19.2 The Magnetic Field of the Earth

Since the north seeking end of a compass points to the opposite polarity (opposites attract), it must be that the magnetic field in the north is really a south pole. Thus we can say:

The magnetic north pole corresponds to the south geographic pole, and the magnetic south pole corresponds to the north geographic pole.

Many living organisms, including some crabs, homing pigeons, and even anaerobic bacteria, have internal compasses and can use the earth's magnetic field to orient themselves and even to navigate during migrations.

The magnetic poles are not aligned with the earth's rotational axis. Also the poles move and occasionally reverse for a couple hundred thousand years or so. It is speculated that this is disorienting to the organisms using the field as a direction indicator. The reversals are recorded in the layers of lava emanating from volcanos and the seafloor on either side of the cracks in the ocean where continental drift is taking place.

The earth's fields also protect us from charged particle cosmic rays and focus the charged particles in the solar wind and from solar flares towards our north and south poles where a beautiful display of northern or southern lights occurs.

19.3 Magnetic Fields

The magnetic field does not affect charges at rest but does have an effect upon charges moving with velocity v according to the following equation

$$F = qvB\sin\theta \qquad (19.1)$$

where B is the magnetic field and has its magnitude defined by this equation. It is measured in units of Tesla, 1 T = (Wb=Webers)/m^2 = 1 N/C-m-s = N/A·m = 10^4 Gauss (cgs unit). The force is always perpendicular to both v and B and is zero when v is parallel to B. *Also, since the magnetic force is always perpendicular to the velocity, it can do no work on the particle.*

19.4 Magnetic Force on a Current-Carrying Conductor

For a current in a conductor we have charges in motion with an average speed of the drift velocity, v_d. The force on the wire, therefore, must be the sum of the forces on the individual charges moving through the wire. Thus, we expect $F = qv_dB \sin\theta\, nAL$, and since the current in the wire is $I = nqv_dA$ and L is parallel to v_d, it simplifies to

$$F = BIL\sin\theta \quad \text{and} \quad F_{max} = BIL \qquad \text{(19.2)}$$

The total magnetic force on an arbitrary shaped wire in an external field **B** would be computed by adding up the effects on many small segments. It is equivalent to the effect on a straight wire oriented from the beginning point to the end point.

19.5 Torque on a Current Loop

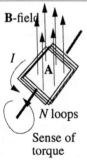

B-field

I

A

N loops

Sense of torque

The torque due to a magnetic field B on a current loop of area A is calculable from our equation of the effect of a magnetic field on a wire of arbitrary shape. The torque is $LFsin\theta$, so the final result is

$$\tau = BIA\sin\theta \quad \text{and} \quad \tau_{max} = BIA \qquad \text{(19.3)}$$

The direction of **A** is perpendicular to the loop in the right-hand sense. If there are N loops then $\tau = NIAB$.

19.6 The Galvanometer and Its Applications

THE GALVANOMETER

The torque produced on a current loop in the presence of a magnetic field can be used to measure the current. The deflection of the loop can be used to measure both currents (ammeters), and voltages (volt meters).

A GALVANOMETER IS THE BASIS OF AN AMMETER

To be used as an ammeter, a galvanometer of very low resistance is needed so that the amount of current flowing in the circuit is essentially unimpeded. The ammeter is put in series with the current that is to be measured. However, even milliampere currents will give large deflections of the needle. Therefore, to avoid overloading the galvanometer or impeding the current to be measured, we must put in a low resistance *shunt* or bypass resistor through which most of the current will flow.

A GALVANOMETER IS THE BASIS OF A VOLT-METER

When a galvanometer is used in parallel with a circuit element and if the internal resistance of the galvanometer is high (100,000 ohms or so), very little current will flow through the galvanometer. Then the deflection due to the current will measure the voltage across the circuit element. Thus a high resistance is placed in series with the galvanometer and the circuit of the resistor and galvanometer is placed in parallel across the circuit element whose potential is to be measured.

19.7 Motion of a Charged Particle in a Magnetic Field

Consider a charged particle in motion in a uniform magnetic field and the velocity of the particle is perpendicular to the field. The force is perpendicular to both the field and velocity and hence is directed towards the center of a circular-orbit trajectory for the particle. We can write down the magnetic force and call it the centripetal force. Thus we get

$$F = qvB = \frac{mv^2}{r} \quad \text{and thus} \quad r = \frac{mv}{qB} \qquad \text{(19. 4)}$$

We see the radius is proportional to the momentum, mv.

We get a spiral orbit (helix) if the velocity is not perpendicular to the magnetic field.

19.8 Magnetic Field of a Long, Straight Wire and Ampere's Law

During (or just after) a lecture in 1819 Hans Oersted found that an electric current produced a deflection in a magnet. Thus he discovered that currents produce magnetic fields.

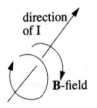

direction of I

B-field

The direction of B around a wire is consistent with the right-hand rule, which is:

If the wire is grasped in the right hand with the thumb in the direction of the current, the fingers will curl in the direction of B.

The strength of the field is given by

$$B = \frac{\mu_0 I}{2\pi r}$$

(19. 5)

where r is the perpendicular distance from the wire to the point of observation and μ_0 is the permeability of free space, a constant defined to have the value $\mu_0 \equiv 4\pi \times 10^{-7}$ (T · m/A)

AMPERE'S LAW

Ampere gave us a scheme for calculating magnetic fields due to currents. If a loop is drawn around some currents, the component of **B** parallel to the tangent of the loop times a short segment can be added up around the loop and the sum is $\mu_0 I$. So mathematically Ampere's law is

$$\sum B_{parallel} \Delta l = \mu_0 I$$

(19. 6)

19.9 The Magnetic Force Between Two Parallel Conductors

The magnitude of the magnetic field around a long straight wire is determined to be $B = \mu_o I/2\pi d$, where d is the distance from the wire to the point where B is wanted. B encircles the wire.

If another long straight wire of length ℓ is placed parallel to the first and at the distance a, then the current in that wire will experience a force due to the B field created by the first wire. Thus the force on wire 1 due to wire 2 is

$$F_1 = I_1 \ell \left(\frac{\mu_0 I_2}{2\pi d} \right) = \frac{\ell \mu_0 I_1 I_2}{2\pi d}$$

(19. 7)

The force on wire 2 due to wire 1 is equal and opposite. This formula can be rewritten so we have the force per unit length, namely

$$\frac{F_1}{\ell} = \frac{\mu_0 I_1 I_2}{2\pi a}$$

(19. 8)

When the wires are 1 meter apart we **operationally define the ampere** as that current in each wire that produces a force of 2×10^{-7} N/m.

19.10 Magnetic Field of a Current Loop

The magnetic field produced by a single, circular loop of wire looks similar to that produced by a short dipole magnet. At the center of the loop the field is $B = \mu_0 I/2R$.

19.11 The Magnetic Field of a Solenoid

A solenoid is a long wire wound in the form of a helix. Tightly wound solenoids produce quite uniform fields inside. Using Ampere's law to determine the field inside we make our summation loop go down the interior, out of the sides of the coil perpendicular to **B** so $B_\parallel \Delta s = 0$ and then out to infinity for the outside edge of the loop where $B = 0$. Therefore, the sum of $B_\parallel \Delta s$ is just $B_{center} \ell = \mu_0 NI$ where N is the number of turns, I

the current in each turn and ℓ the length of the solenoid, that is,

$$B = \mu_0 \frac{N}{\ell} I = \mu_0 nI \qquad \text{(19.9)}$$

19.12 Magnetic Domains

Magnetism in matter is due to electrons in orbit around nuclei in atoms and their intrinsic spin. The current produced by an electron in orbit and by its spin creates a magnetic field. Usually the magnetic field of one electron in an atom cancels the magnetic field of another so there is no net magnetism.

Iron, cobalt nickel, gadolinium and dysprosium are materials in which the magnetic moments of the electrons do not cancel. They are said to be ferromagnetic and are used to make permanent magnets. They contain spin magnetic moments that tend to align parallel to each other. Hence, these materials have magnetic domains and magnetism is achieved because energetically favorable domains tend to grow in size. In certain materials, called hard magnetic materials, the domains remain aligned even after the external field has been removed thus giving permanent magnetism.

19.13 Concept Statements and Questions

1. A magnetic field, **B**, exerts a force on a moving charge that is perpendicular to **B** and **v**, the velocity of the charge. As such, this force is centripetal and cannot

do work on the charge. The **B** field also exerts forces on currents.

2. A moving charge creates a magnetic field. The current in a wire, therefore, creates a magnetic field that encircles the wire like concentric imaginary cylinders. This magnetic field will interact with any external field. Thus a current interacts with a magnetic field. How do you compute the force?

3. Two current loops can mutually interact with each other through their magnetic fields. A specific example of two currents interacting is two long straight current carrying wires. Like (parallel) currents attract and anti-parallel currents repel.

5. Solenoids can be used to create nearly uniform magnetic fields inside the coil. They have a field outside which looks very much like a dipole magnet.

6. How can a magnetic field produce a torque on a current loop?

7. What is the definition of an ampere in terms of forces between wires?

8. What are magnetic domains? Permanent magnetism in a material is achieved in terms of domains?

19.14 Hints for Solving the Problems

General Hints

1. Remember that the magnetic field cannot do work on a free particle moving in the field.

2. Use Ampere's law to get magnetic fields when you have a symmetrical situation.

3. Use the right-hand rule to get the direction of **B** due to current.

Hints for Solving Selected Problems

1-12. The magnetic force formula works on these problems. The magnetic force is centripetal and, therefore, $F = ma_c$ applies.

13-23. Use the equation for a force on a wire due to a magnetic field. Keep track of the direction for each segment of the wire, and combine vectorially when a sum or net affect is asked for. Levitation is possible when there is a balance between the gravitational force with the *BIL* force.

24-26. Use the equation for a torque due to a magnetic field on a current loop. If the magnetic moment is involved it can be calculated from IA and vice-versa.

27-32. Study Section 19.6.

33-50. Equate the magnetic force to the formula for centripetal force ($F = mv^2/r$) and also from conservation of energy the kinetic and potential energies can be numerically equal so $mv^2/2 = qV$. Use these equations together. Ampere's law may be useful on 49.

51-54. Use the equation for force or force per unit length between two long parallel conductors.

20
Induced Voltages and Inductance

**Electric Fields
and Voltages
from Changing
Magnetic Fields**

Magnetic fields not only produce a force on a moving charge, a changing magnetic field also produces an electric field. Faraday's law of induction relates magnetic flux change to the emf produced, a process called *induction.*

Inductors are circuit elements that produce an emf that **opposes** the external emf causing the change. This is an effect known as Lenz's law and a process of self-inductance.

20.1 Induced emf and Magnetic Flux

It is easily demonstrated that when current changes occur in a coil, a voltage or emf is produced in a nearby coil. This "induced" emf is produced because the primary coil creates a changing magnetic field through its changing current. To better understand this we need to first discuss magnetic flux.

MAGNETIC FLUX

Magnetic flux, Φ, is the product of the magnetic field, **B**, and the effective area (area perpendicular to the field) through which the field passes, so

$$\Phi = B_\perp A = BA_\perp = BA \cos \theta \qquad \text{(20. 1)}$$

The SI unit of flux is tesla-meter^2, which is named weber (Wb), so $1 \text{ Wb} = 1 \text{ T} \cdot \text{m}^2$.

If lines are drawn to represent the field, the density of the lines (number per unit cross-sectional area) will be proportional to the field strength and the total number will be proportional to the flux.

20.2 Faraday's Law of Induction

Faraday found that the relative motion between a magnetic field and a coil of wire produces an electric current without the aid of a battery. We call such a current *an induced current,* and the emf that produced it *an induced emf.*

Faraday's law states: ***The emf induced in a circuit is directly proportional to the time rate of change of magnetic flux through the circuit***. This can be written

$$\xi = -\frac{\Delta \Phi_m}{\Delta t} \qquad \text{(20. 2)}$$

A coil with N loops will produce N times as much emf, and for a uniform magnetic field the induced emf is

$$\xi = -\frac{\Delta (BA \cos \theta)}{\Delta t} \qquad \text{(20. 3)}$$

The minus sign in this equation and the general sense of the outcome can be stated as Lenz's law, which says

The polarity of the induced emf is such that it produces a current whose magnetic field opposes the change in magnetic flux through the loop. That is, the induced current tends to maintain the original flux through the circuit.

20.3 Motional emf

A straight conductor of length l which is moving so it cuts across magnetic field lines can be thought of as a line of positive and negative charges moving through a magnetic field. So the charges experience a magnetic force which sends the negative charges to one end of the wire and the positive charges to the other end. This produces an emf between the two ends and the charges stop flowing when the Coulomb force, qE, equals the magnetic force, qvB *(for $\theta = 90^o$)*. The voltage V or emf between the two ends is El. Thus,

$$V = El = Blv \qquad (20.4)$$

and the emf continues as long as the relative motion between the wire and **B** field continue. If we connect the two ends of the wire with a conductor, we get a current flow.

Once we have a current loop, we can view this experiment as magnetic flux changing inside the loop due to the motion of one side of the loop which increases the area. If the area is lx, where x is the variable side of the rectangle loop and $v = \Delta x/\Delta t$, then the emf induced is

$$\xi = -\frac{\Delta \Phi_m}{\Delta t} = -\frac{\Delta}{\Delta t}(Blx) = Blv \qquad \text{(20. 5)}$$

The current produced will equal the emf divided by the internal resistance R.

20.4 Lenz's Law Revisited

When the magnetic flux is changed due to the motion of part of a circuit, for example, enlarging the loop over a uniform magnetic field, or decreasing the loop's size, this induces a current which opposes the change (increase or decrease). Any source of change of flux is opposed so as to maintain the status quo, as much as possible, as far as flux is concerned.

20.5 Generators

A coil of wire rotating in a uniform magnetic field will produce alternating emf across the coil. If the coil is connected to an outside circuit we find an alternating current. Such a device for producing electricity is called a generator. The power to rotate the coil comes from elsewhere. Running this device backwards, i.e. inputting the alternating current and emf, creates a magnetic field in the coil which causes the coil to turn and from such a rotation, external power may be obtained. We call this device an electric motor.

A direct-current generator is possible by switching the way the current is taken off every half cycle with a split-ring commutator. Similarly DC motors can be made. The ripples in the current and emf can be smoothed out with a properly designed filter circuit. The emf produced by the AC generator is given by

$$\xi = NBA\omega\sin\omega t \quad \text{and} \quad \xi_{max} = NBA\omega \qquad \text{(20. 6)}$$

MOTORS AND BACK emf

In accordance with Lenz's law, when the coil of the motor rotates in a magnetic field, it creates an opposing or back emf. The back emf reduces the voltage supplied by the power source, which enters the motor and reduces the power utilized. The back emf is smallest when the motor is doing the most work and drawing the most power.

20.6 Eddy Currents

Any conductor in the presence of changing magnetic fields will experience small circulating currents called eddy currents due to the motion of the free charges that are caused to move by the changing field. Eddy currents in large flat plates can circulate over large regions of the plate, whereas if the conductor is in small segments or strips, the eddy currents are reduced.

The eddy currents can be viewed as a response by the conductor to Lenz's law to create a magnetic flux in opposition to the change of flux occurring in the proximity of the system.

Eddy currents are usually undesirable because of the loss of energy in an electrical system due to their presence, so considerable engineering effort is devoted to designing systems that have minimal eddy currents.

20.7 Self-Inductance

The changing magnetic field due to changing currents in one part of a circuit can interact with those produced in another part of the circuit. According to Lenz's law the effect always opposes the imposed changes. Thus we get self-induced effects. The self-induced emf is always proportional to the time-rate of change of the current. Thus, a coil in a circuit with closely spaced turns will give an emf as

$$\xi = -N\frac{\Delta\Phi_m}{\Delta t} = -L\frac{\Delta I}{\Delta t}$$

(20. 7)

L is called the inductance of the coil. It depends on the geometric features of the coil or circuit and is calculable from

$$L = \frac{N\Phi_m}{I}$$

(20. 8)

Inductance is a measure of opposition to a changing current just as resistance is opposition to current. The unit of inductance is the henry (H), so $1\ H = V{\cdot}s/A$.

20.8 RL Circuits

Applying Kirchhoff's loop rule as we go around a circuit with both a resistor and an inductor we get

$$\xi - IR - L\frac{\Delta I}{\Delta t} = 0$$

(20. 9)

which is an equation for the current I. The first term, IR, indicates a resistance to current, whereas the second term, $L\Delta I/\Delta t$, indicates an "opposition" to a changing current. Together the two effects give a gradual increase in current such that

$$I = \frac{\xi}{R}(1 - e^{-Rt/L}) = \frac{\xi}{R}(1 - e^{-t/\tau}) \qquad \text{(20. 10)}$$

Here $\tau = R/L$ is the relaxation time or the time for the current to reach $(1-e^{-1}) = 0.63$ of its final value.

20.9 Energy Stored in a Magnetic Field

The rate at which energy is stored in an inductor is a product of the current and the emf of the inductor. The amount of energy stored at any instant is given by

$$PE_L = \frac{1}{2}LI^2 \qquad \text{(20. 11)}$$

20.10 Concept Statements and Questions

1. Changing magnetic flux through a loop creates an emf in the loop. If the loop is a piece of conducting wire, will this result in a current?

2. Is it possible, according to Faraday's law, to generate a current by moving a magnet back and forth inside a conducting coil? How? Would such a device be called an electrical generator?

3. How does increasing the number of turns in the coil affect the emf produced by a changing B-field?

4. An emf is produced inside a loop of wire whenever the number of magnetic flux lines changes through the loop regardless of what produces the change.

5. To change the amount of magnetic flux passing through a wire loop requires work or power. This power is dissipated through heating of the wire due to its resistance. If the wire is a superconductor, the energy will go into the energy of the magnetic field produced around the superconducting current.

6. Two coils are co-aligned but are not touching. They are wound in the same sense. If a changing current is decreasing in the first coil, will the induced current in the second coil be in the same sense or opposite? What does Lenz's law say?

7. The inductance of a coil, in air, can be given entirely in terms of geometry factors.

9. What inhibits the flow of current when charging an inductor?

20.11 Hints for Solving the Problems

General Hints

1. Lenz's law is simple but often confusing. Rehearse it and practice applying it. It is used in many problems and once mastered is very helpful.

2. Don't forget the minus sign in Faraday's law when using it in problems.

Hints for Solving Selected Problems

1-22. For Faraday's law to be applied you need the changing magnetic flux so you need to be able to calculate flux. It is the amount of magnetic field lines passing through an area (the area of the loop). If the magnetic field is a function of time, you can get $\Delta B/\Delta t$ for your changing flux, or the area may change. Any of these changes will produce an emf.

22-29. Use Blv for the changing flux for conductors in motion relative to the B-field to generate an emf. In rotating coils, the volt meter or ammeter will fluctuate if the rotation is slow. Otherwise, you need to calculate the average, which is the average of the cosine function over a half-cycle

30-38. These problems give you practice using Lenz's law. Remember, it is the change which is opposed not the flux or current.

39-47. The frequency of an alternating generator is obtained from ω as $f = \omega/2\pi$.

48-53. Use the self-inductance equation for relating changing current and inductance to emf. Inductance may be calculated from the flux, number of turns and current, i.e. $L = N\Phi_m/I$. Look at Example 20.8

54-58. You will find Equation 20.10 in this Pocket Guide to help you with the change of currents at different times.

21
Alternating Current Circuits and Electromagnetic Waves

Waves that Create Each Other

Alternating currents (ac) are found in our homes and our work places and have become a part of every day life. We need to understand the basic physics associated with them. To do this we need to know how resistors, capacitors, and inductors work in ac circuits.

21.1 Resistors in an ac Circuit

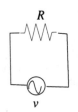

A power source for ac voltage and current could be an ac generator. The current and voltage will be in phase with each other and we can represent them by

$$v = V_m \sin(2\pi ft) \quad \text{and} \quad i = I_m \sin(2\pi ft) \qquad (21.1)$$

Here f is the frequency of the alternating current power source and v and i are the instantaneous voltage and current measured in volts and amperes, respectively.

The power consumed by the circuit is given by

$$P = iv = i^2R = v^2/R \qquad \text{(21.2)}$$

Here we have used the Ohm's law relationship, that $V = IR$, as it applies to the resistor R.

The total current over a complete cycle is zero, but the average magnitude of the current is not and is obtained by taking the square root of the average of the square. This is called the root mean square or rms value. Since the sine squared over a cycle is 1/2 we get

$$I_{rms} = \frac{I_m}{\sqrt{2}} = 0.707 I_m \quad \text{and} \quad V_{rms} = \frac{V_m}{\sqrt{2}} = 0.707 V_m \quad \text{(21.3)}$$

21.2 Capacitors in an ac Circuit

There is a current into and out of a capacitor as it is charged. When the voltage achieves line voltage the charging ceases and there is no longer a current. Reversal of the direction of the current due to ac voltages once again allows a current in the opposite direction. This switching back and forth provides an impedance, (X_C measured in ohms), to flow which depends on the frequency, f, according to

$$X_C \equiv \frac{1}{2\pi f C} \qquad \text{(21.4)}$$

The rms voltage is related to the rms current in a formula similar to Ohm's law for capacitors in ac circuits so that

$$V_C = IX_C \qquad \text{(21.5)}$$

The voltage lags the current by $90°$ across a capacitor.

21.3 Inductors in an ac Circuit

If we put an inductive coil into an ac circuit, we know the voltage across the coil is given by

$$V_L = IX_L \qquad \text{(21.6)}$$

where the current is impeded by the back emf of the coil which is frequency dependent according to

$$X_L \equiv 2\pi f L = \omega L \qquad \text{(21.7)}$$

The voltage leads the current across an inductor by 90°.

21.4 The RLC Series Circuit

Suppose we have an ac circuit with $i = I_m \sin 2\pi f t$ which contains a resistor, a capacitor, and an inductor.

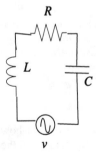

We know the voltage in the resistor is in phase with the current but in the capacitor it lags by 90° and in the inductor it leads by 90°. What is the phase between all three? The answer is obtained by combining the three voltages as voltage phasors with V_R along the x-axis and V_L along +y and V_C along -y so that the resultant voltage and phase are given by

$$V = \sqrt{V_R^2 + (V_L - V_C)^2} \qquad \tan\Phi = \frac{V_L - V_C}{V_R} \qquad \text{(21.8)}$$

Written in terms of impedance the resultant voltage and phase are

$$V = IZ \quad \text{where} \quad Z \equiv \sqrt{R^2 + (X_L - X_C)^2}$$

$$\text{and} \quad \tan\phi = \frac{X_L - X_C}{R} \tag{21.9}$$

21.5 Power in an ac Circuit

Power in ac circuits is dissipated only by resistors and that power shows up as heat. No power is lost in the capacitor nor in the inductor because the current and voltage are 90° out of phase. The average power dissipated, therefore, is

$$P_{av} = I^2 R = IV_R = IV\cos\varphi \tag{21.10}$$

The quantity $cos\ \phi$ is called the power factor.

21.6 Resonance in a Series RLC Circuit

From the relationship $I = V/Z$, it is clear that I is greatest whenever Z is smallest and that occurs when $(X_L - X_C) = 0$. The total reactance is zero when the inductive and capacitive reactance are equal and so

$$2\pi f_0 L = \frac{1}{2\pi f_0 C} \quad \text{or} \quad f_0 = \frac{1}{2\pi\sqrt{LC}}. \tag{21.11}$$

The receiving circuit of a radio is an example of how the strongest signal is received and amplified when the frequency is tuned so that the reactance is zero.

21.7 The Transformer

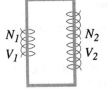

Transformers are used to increase (step-up) the voltage in a circuit with a simultaneous reduction or step-down of the current or a decrease (step-down) of the voltage and increase of the current. This is done with two coils, the primary coil and the secondary coil, which affect each other through inductance. The maximum power is achieved with a design that includes a soft iron core that passes through both coils and reconnects onto itself so as to trap all of the magnetic flux. The voltage in each coil is proportional to the number of turns in each and the rate of change of the flux, but since the rate of change of flux in each is the same we get

$$V_2 = \frac{N_2}{N_1} V_1 \qquad (21.12)$$

In an ideal transformer the power output equals the power input so we can write

$$I_1 V_1 = I_2 V_2 \qquad (21.13)$$

21.8 Maxwell's Predictions

James Clerk Maxwell (1831-1879) was able to take the electric and magnetic phenomena we have talked about, unite them into a single theory of electromagnetism and predict that light was an electromagnetic wave which traveled at the same speed in a vacuum as all other electromagnetic waves regardless of wavelength. His theory was based on the following four pieces of information:

1. Electric fields originate on positive charges and terminate on negative charges. The electric field due to a point charge can be determined at a location by applying Coulomb's force law to a test charge placed at that location.

2. Magnetic field lines always form closed loops; that is, they do not begin or end anywhere.

3. A varying magnetic field induces an emf and hence an electric field. (Faraday's law)

4. Magnetic fields are generated by moving charges (or currents), as summarized in Ampere's law.

Many of the phenomena associated with electromagnetic waves can be understood and described with Maxwell's theory. Some, such as the particle properties of light, cannot.

21.9 Hertz's Discoveries

Heinrich Hertz in 1887 was able to demonstrate the existence of electromagnetic waves produced by an LC circuit that traveled at the speed of light ($c = 3 \times 10^8$ m/s) over a distance of several meters, and then detect them with another LC circuit called the receiver tuned to the same frequency,

$$f_0 = (2\pi\sqrt{LC})^{-1} \qquad \text{(21.14)}$$

The circuit is analogous to a mass oscillating on the end of a spring.

Hertz knew the frequency of the wave and doing an interference experiment he was able to measure the wavelength λ. The product of f and λ gives the speed of the wave and in this way Hertz showed the speed to be 3×10^8 m/s, the speed of light.

These discoveries greatly supported Maxwell's theory.

21.10 The Production of Electromagnetic Waves by an Antenna

Electromagnetic (EM) waves are emitted by any circuit carrying an alternating charge. Any charged particle which accelerates will either radiate or absorb EM radiation. Therefore, an alternating voltage attached to an antenna forces charges in the antenna to oscillate and, therefore, radiate EM waves. A radio station will use such a set-up to broadcast its program. The waves can be picked up by a receiving antenna far away. The oscillating electric field in the wave causes charged particles in the antenna to oscillate in the same way. The weak signal can be amplified so as to exactly match the oscillations in the passing wave.

21.11 Properties of Electromagnetic Waves.

Even though the waves become weak when traveling long distances, the changing electric field always produces a magnetic field in phase but perpendicular to it and to the direction of propagation and vice versa.

The speed of the wave is $c = 1/\sqrt{\mu_0 \varepsilon_0}$ where $\mu_0 = 4\pi \times 10^{-7}$ Wb/A-m and $\varepsilon_0 = 8.85418 \times 10^{-12}$ C^2/N-m^2 giving for the value of c, 2.99792×10^8 m/s.

The ratio of the electric to the magnetic field is the speed of the wave in that medium. For a vacuum, that is

$$\frac{E_{max}}{B_{max}} = \frac{E}{B} = c, \quad \text{or} \quad E = cB \qquad (21.15)$$

The rate of flow of energy/unit area, S_{av} = the average power/area in an electromagnetic wave, is described by the expression

$$S_{av} = \frac{E_{max} B_{max}}{2\mu_0} = \frac{E^2_{max}}{2\mu_0 c} = \frac{c}{2\mu_0} B^2_{max} \qquad (21.16)$$

Electromagnetic waves also carry momentum and, therefore, can exert pressure on a surface by absorption and/or reflection.

21.12 The Spectrum of Electromagnetic Waves

The range of wavelengths for electromagnetic waves goes from zero to infinity. They all travel with the speed of light through the vacuum and follow the equations we have developed in this chapter. We have given names to specific ranges which go approximately as follows.

Radio (> 30 cm), Microwave (1 mm - 30 cm), Infrared (7×10^{-7} m -1 mm), Visible (4×10^{-7} - 7×10^{-7} m),

Ultraviolet (3.8×10^{-7} - 6×10^{-8} m), X-Rays (10^{-8} - 10^{-13} m), and Gamma rays ($< 10^{-13}$ m).

21.13 Concept Statements and Questions

1. Resistive ac circuits have the current and voltage in phase, whereas in capacitive circuits the voltage lags the current and in inductive circuits it leads the current by 90^0.

2. The impedance to current in capacitive circuits is called capacitive reactance, X_C. How is it calculated? What is the impedance produced by an inductance called and how is it calculated?

3. How does one determine the total impedance to current in a series RLC circuit? How is the phase angle between the voltage and current determined?

4. The power, P, in an RLC circuit is given by _____? where $\cos \phi$ is called the power factor.

5. The natural frequency of an RLC circuit is the same as the resonance frequency. How is f_0 determined?

6. A transformer works on the principle of inductance. Given the ratio of the number of turns of wire on each of the two coils of a transformer, how does one determine the ratio of the voltages? Can the ratio of currents also be obtained? How?

7. All of the properties of electromagnetic waves are derivable from Maxwell's four equations which are

obtained from the four principles stated. Hertz verified the correctness of Maxwell's theory.

8. The electromagnetic wave has oscillating **E** and **B** fields that are transverse. They are perpendicular to each other and perpendicular to the direction of travel. Also, $E = cB$.

9. Light, x-rays, radio-waves, gamma rays, infra-red rays, and ultraviolet rays are all electromagnetic waves that travel at 3×10^8 m/s and differ one from the other only by wavelength or frequency.

21.14 Hints for Solving the Problems

General Hints

1. You will use $E = cB$ many times, but it is true only for EM radiation, so watch out.

2. Distinguish carefully between rms values, average values and maximum values for voltage, current, and power.

3. While in many respects reactance and resistance are similar, when they are combined to give impedance you must consider the $90°$ phase difference and use the impedance formula.

4. Refer to and study the examples in the text. Also the Problem-Solving Strategies in Section 21.4 are very helpful.

4. Be careful not to mix-up momentum of an EM wave and the pressure this momentum creates when it hits a surface.

Hints for Solving Selected Problems

1-5. Refer to the formulas for average power in Section 21.5 and use Ohm's law ($V = IR$) as needed in some of the problems.

7-10. Compute capacitive reactance from Equation 21.4, and from that you can get current or voltage if the other is given from Equation 21.5.

11-16. Look at Section 21.3 and the equations there.

17-27. Compute the impedance in the circuit and use it to determine the current and voltage in the circuits.

28-32, You will need the phase angle in addition to the voltage and current to compute the power.

33-40. In a series *RLC* ac circuit, the inductive and capacitive reactance disappear at the resonance frequency. The impedance is thus minimized and the largest currents are possible for a given voltage. See Section 21.6 and Example 21.6.

41-43. Use the transformer equations.

44-45. The tuning frequency is the same thing as resonance frequency. With L fixed, the variability in C gives variability in the frequency.

46-54. Use Equations 21.15 and 21.16 and $E = cB$.

55-61. Use $f\lambda = c$ to convert between frequencies and wavelengths and $c = d/t$ to get times and distances.

22
Reflection and Refraction of Light

**Light and EM
Waves at
Boundaries**

In the previous chapter light was presented as being a part of the electromagnetic spectrum. Long before it was understood as an EM phenomenon it had been studied and known to be dispersed into many colors, had a very high speed, would reflect and refract, would interfere like other waves, and could be polarized. We study some of these attributes in this chapter.

22.1 The Nature of Light

Light has been demonstrated to be waves. Thomas Young showed this in 1801 by interference of two coherent waves. Before being shown that it was waves, scientists believed it was made of particles. Today it is known to be made of particles called photons and it is also waves. The energy of the photon is proportional to its frequency and all photons of a given frequency have the same amount of energy.

$$E = hf \qquad \text{(22· 17)}$$

The constant h is known as Planck's constant and is the same h used in describing the angular momentum and spin on the atomic level. Its SI value is 6.63×10^{-34} J-s.

22.2 Measurements of the Speed of Light

Early experiments to measure the speed of light failed because the speed was so high.

ROEMER'S METHOD

By timing the moments of occultation of Jupiter's moons by the giant planet, Ole Roemer observed that the further away the earth got on its orbit from Jupiter, the later they occurred. He assumed the extra time required was due to the travel time of the light over the extra distance. He was right and was able to measure the speed of light to an accuracy of about 2/3 of its presently known value.

FIZEAU'S TECHNIQUE

Armand Fizeau passed a beam of light through the gaps between teeth of a rapidly rotating wheel and detected the reflected beam when it again passed the wheel. At the fastest rotation rate for light to be returned, the reflected beam passed through a gap adjacent to the original gap. From the distance traveled and the speed of rotation of the wheel, the speed of light was calculated within about 3%.

22.3 Huygens' Principle

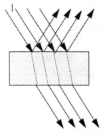

Huygens assumed light is waves and devised ways of understanding the way these waves would move. He devised a rule for drawing the propagation of light waves known as Huygens' principle. The rule is,

All points on a given wavefront are taken as point sources for the production of spherical secondary waves, called wavelets, that propagate outward with speeds characteristic of waves in that medium. After some time has elapsed, the new position of the wavefront is the surface tangent to the wavelets.

THE RAY APPROXIMATION IN GEOMETRIC OPTICS

Light rays are parallel lines drawn along the direction of travel of light and represent a beam of light. As such, they are perpendicular to the wave front rendition of light that moves out from a source like a water wave produced by a pebble. Both of these representations are only approximations and need careful modification and clarification when the light passes close to an edge or goes through a small opening so that diffraction effects become important.

22.4 Reflection and Refraction

REFLECTION OF LIGHT

As parallel light rays are incident upon a flat reflective surface, they reflect so as to make the angle of reflection equal to the angle of incidence. Both angles are measured between the ray and a line drawn perpendicular to the surface. Thus, the law of reflection is

$$\theta_i = \theta_r \qquad \text{(22· 18)}$$

REFRACTION OF LIGHT

In passing through a flat boundary and continuing on the other side into a medium where the speed is different, a light ray will change its direction according to Snell's law given as

$$\frac{\sin\theta_2}{\sin\theta_1} = \frac{v_2}{v_1} = \text{constant} \qquad \text{(22· 19)}$$

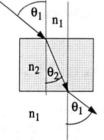

The path of a light ray through a refracting surface is reversible. The path taken is independent of the direction of travel of the wave.

Light usually travels more slowly in a medium where there are atoms and electrons than in a vacuum. The change in speed is caused by a phase shift when the wave interacts with particles of the medium.

22.5 The Law of Refraction

It is convenient to define the index of refraction as

$$n \equiv \frac{\text{speed of light in vacuum}}{\text{speed of light in a medium}} = \frac{c}{v} \qquad \text{(22· 20)}$$

In passing from one medium to another, the frequency remains constant. Since the speed changes, however, the wavelength must also change. The connection between wavelength and index of refraction can be written and Snell's law of refraction can also be written in terms of index of refraction for each side of a boundary.

$$\lambda_1 n_1 = \lambda_2 n_2, \quad n = \frac{\lambda_0}{\lambda_n}$$

$$n_1 \sin\theta_1 = n_2 \sin\theta_2$$

(22-21)

22.6 Dispersion and Prisms

Rays of different wavelengths travel at different speeds in materials, therefore the index of refraction is a function of wavelength. Accordingly, the angle of refraction will be different for each wavelength. The shorter the wavelength the more it interacts and the slower it goes. Therefore, blue light is refracted (bent) more in going into a denser medium than red light (red wavelength ~ twice as long). This spreading of colors is called *dispersion*. With a prism light can be dispersed at each face, increasing the dispersion. This prism effect is well known to produce a beautiful color spectrum of visible light. Remember, blue light is bent the most.

22.7 The Rainbow

The rainbow is a good example of the dispersion of light. The light enters the droplets of water, refracts upon entering and reflects internally and finally re-emerges. Because different wavelengths are refracted by different amounts the rays become dispersed and emerge at different angles. An observer sees differing colors dependent upon the direction of observation.

22.8 Huygens' Principle Applied to Reflection and Refraction

Huygens' principle can be used successfully to construct the law of reflection because the reflected wavelets travel the same distance in the same amount of time giving the angle of reflection to be the same as the angle of incidence. In refraction, however, the refracted wavelets travel more slowly and this changes the direction of the wavefront (which is perpendicular to the light rays) thus we get Snell's law, $n_1 sin\ \theta_1 = n_2 sin\ \theta_2$.

22.9 Total Internal Reflection

Since the path of light rays is reversible we can understand how total internal reflection occurs. That is, if a light ray is inside a denser medium and is incident at an angle great enough so the exit angle is 90°, there can be no such refracted ray and all of the intensity appears in the reflected ray. This angle of incidence is called the critical angle and is given by The critical angle is given by

$$n_1 \sin\theta_1 = n_2 \sin(\pi/2) \quad so \quad \sin\theta_c = \frac{n_2}{n_1} \qquad \text{(22• 22)}$$

Total reflection can only occur when a light ray goes from a medium with a higher index of refraction to one that is lower.

FIBER OPTICS

A light beam can travel from one end of a transparent fiber to the other even though the fiber bends and curves because the beam always hits the surface at an angle greater than the critical angle and is totally internally reflected. There are a few light rays that scatter at a smaller angle and, thus, get out.

22.10 Concept Statements and Questions

1. Light has both wave-like and particle-like properties. Which is it?

2. Would Roemer's method for measuring the speed of light work for Saturn and its moons?

3. Huygens' principle treats each point on a wavefront as a source of a new wave.

4. Light reflects at the same angle it came in, the angle of incidence. This can be shown to be true whether light is a particle or a wave.

5. The law of refraction of light is a consequence of light traveling at a lower average speed in a medium than in the vacuum.

6. When the angle of incidence is almost 90°, the resulting angle of refraction is called the critical angle. For angles greater than the critical angle and inside the medium, total reflection will occur at the boundary.

7. Does the frequency of light change in going from one medium to another? How about the wavelength?

5. What is Huygens' principle and what is it used for? Can it be used on sound waves and water waves as well as light? (Yes! It works for any wave.)

6. Dispersion occurs because n depends on wavelength. Which color is bent the most in going from a lower to a higher index of refraction?

22.11 Hints for Solving the Problems

General Hints

1. To use Snell's law correctly it is necessary to know what happens to the index of refraction, or equivalently the speed of light in the two media. Study the relationship and make sure you understand it.

2. The wavelength of a light wave is always inversely proportional to the index of refraction or directly proportional to the speed of the wave in that medium. (Do water waves also refract? Yes! If there is a boundary where the depth of the water changes, the speed of the wave changes and the wave refracts.)

3. Study the examples.

Hints for Solving Selected Problems

1-5. Remember angular speed is given as $\omega = \Delta\theta/\Delta t$ whereas linear speed, v, is $\Delta x/\Delta t$. You may need these expressions in these problems.

6-8. Make sketches of the known geometry and light rays. The sketches help you to understand the problem and what you are solving for.

9-12. Use the definitions and laws of reflection and refraction. These problems mostly use the definition of index of refraction.

13-31. Use the law of reflection, Snell's law, geometry and/or trigonometry on most of these problems. A few require the use of the effect of index of refraction on wavelength.

32-35. Use the geometry of prisms and Snell's law.

36-48. Use the formula and definition of critical angle for these problems. For computing the critical angle between two media, it is the relative index of refraction (n_1/n_2) that counts.

23
Mirrors and Lenses

**Reflection and
Refraction at
Shaped
Boundaries**

The process of image formation is extremely useful
and practical as we consider telescopes, microscopes,
eye glasses, projectors, magnifiers, etc. These involve
lenses which interact with spherical wavefronts coming from point sources and reshape them to come to
focus at another point, and thus make images. Both
curved mirrors and curved refracting surfaces can be
used for this purpose. This chapter deals with some of
the simple concepts involved in creating images.

23.1 Plane Mirrors

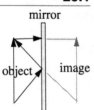

Because of the law of reflection, light from an object
striking a plane mirror reflects to give an image inside
the mirror that is virtual since the light rays don't actually go or come from there. They only appear to come
from there, hence the image is virtual.The image also
will be erect, the magnification (image height/object
height) will be 1, and it will be the same distance from

the surface (but inside) as the object is outside. The image will have a perceived left-right reversal but actually there is a front-back reversal instead. A persons image of a right hand will be a reflection of his left hand, and vice-versa. The magnification for any mirror or lens system is defined as

$$M \equiv \frac{\text{image height}}{\text{object height}} = \frac{h'}{h} \tag{23.1}$$

23.2 Images Formed by Spherical Mirrors

The surface of a spherical mirror has a spherical shape, at least a segment of a sphere. It does not focus precisely because a parabolic surface is needed for that, but it approximates a parabolic surface fairly well and is much easier to make.

CONCAVE MIRRORS

If the object distance from the lens is p and the image distance is q, then

$$\frac{1}{p} + \frac{1}{q} = \frac{2}{R} = \frac{1}{f} \tag{23.2}$$

where R is the radius of curvature and f is the focal length of the lens. If p is positive when to the left, then positive q is also to the left. Also f is positive. If q is positive the image is real, if q is negative it is virtual.

The magnification which is the image height divided by the object height is also given by $M = -q/p$.

23.3 Convex Mirrors and Sign Conventions

The same lens formula applies as for the concave case but now R and hence f are negative. p is positive to the left, q negative is to the right and makes a virtual image since the light rays only appear to come to a focus on the right but cannot actually come from there.

Ray Diagrams for Mirrors

Using ray drawing one can construct the image for mirrors using four rays drawn as follows.

1. Ray 1 is drawn parallel to the principal axis and is reflected back through the focal point, F.

2. Ray 2 is drawn through the focal point. Thus, it is reflected back parallel to the principal axis.

3. Ray 3 is drawn through the center of curvature, C, and is reflected back on itself.

4. Ray 4 is drawn through the center of curvature (at R) and is reflected back on itself.

The intersection of any two of these rays locates the image with the third and fourth rays serving as checks.

23.4 Images Formed by Refraction

For refraction through a single curved surface and going from index of refraction n_1 to index of refraction n_2, the formula is (see the text for the derivation),

$$\frac{n_1}{p} + \frac{n_2}{q} = \frac{n_2 - n_1}{R}$$

(23. 3)

PLANE REFRACTING SURFACES

For a plane surface, R is infinity so the term on the right side of the equation above goes to zero and we have

$$q = -(n_2/n_1)\,p$$

(23. 4)

23.5 Atmospheric Refraction

There are two naturally occurring results due to refraction in the earth's atmosphere. First we have the visible sun after it has physically set, at sundown. It can be seen because atmospheric refraction bends the rays down to us. Second, the mirage produced by hot air near the ground making it possible to see objects above the horizon as though they were below the horizon. In the most common example, we see an image of the sky as though it had reflected off "water" on a hot, dry highway.

23.6 Thin Lenses

R_1 R_2

For thin lenses we have two radii of curvature and they are close enough together so that we can ignore the thickness of the lens. We apply the single curvature formula twice and get

$$\frac{1}{p} + \frac{1}{q} = \frac{1}{f} = (n-1)\left(\frac{1}{R_1} - \frac{1}{R_2}\right)$$

(23. 5)

A converging lens will be thicker at the center than on the edges and will have a positive focal length f. A diverging lens will be thicker at the edges and has a negative focal length. If the rays come from the left then p is positive if on the left side of the lens and q is positive if on the right, otherwise they are negative. The focal length of a converging lens is positive and of a diverging lens negative.

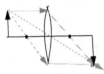

RAY DIAGRAMS FOR THIN LENSES

Three rays can readily be drawn for thin lenses and any two will suffice to determine the position of the image. The three rays are:

1. The first ray is drawn parallel to the principal axis. After being refracted by the lens, this ray passes through (or appears to come from) one of the focal points.

2. The second ray is drawn through the center of the lens. This ray continues in a straight line.

3. The third ray is drawn through the focal point, F, and emerges from the lens parallel to the principal axis.

COMBINATION OF THIN LENSES

For a combination of thin lenses the thin lens formula is applied over and over. The image distance from the first lens is determined, then with that position known, its distance to the next lens is used as the new object distance and a new image is formed, etc. This process is repeated until the final image position is known.

23.7 Lens Aberrations

There are several imperfections in lenses preventing perfect images. Two important ones are **spherical aberrations** and **chromatic aberrations**. Spherical aberrations are a consequence of a spherical shape not being able to give a perfect focus. We could try other shapes but they won't work either. Spherical surfaces are the most easy to make and they work quite well. Chromatic aberrations are due to dispersion, i.e., because the index of refraction is different for each of the colors (wavelengths) then the focal length is differs for each and hence so does the image position. The variations are small, but leave fringes of color around the image. Both of these aberrations can be corrected to a certain degree. In a mirror, a parabolic shaped surface totally eliminates spherical aberration for point sources and images. Reflective surfaces have no chromatic aberration since all wavelengths reflect the same.

23.8 Concept Statements and Questions

1. Lenses and mirrors are used to make images of objects. When the light rays from the object actually come together to form an image, the image is real, but if they only appear to have come together (because they are actually diverging), the image is virtual.

2. The optical system (mirror or lens) can form inverted images or erect images, enlarged or reduced, real or virtual images, depending on the focal length and the position of the object.

3. The rules for drawing ray diagrams for mirrors and lenses to construct images are the same except the images for the mirror are on the opposite side as though they had been reflected (which they have).

4. Combinations of lenses can be worked by taking one lens at a time. Start with the position of the object and use the image produced by the first lens as the object of the second lens and etc.

5. How can the magnification be obtained if the image and object distances from the lens (or mirror) are known?

23.9 Hints for Solving the Problems

General Hints

1. The mirror formula and thin lens formula are identical in appearance. The only difference is the interpretation of the sign of the focal length and the sign of the image distance. The mirror is in a sense a reflection of the thin lens.

2. The hardest part of these problems is keeping the signs straight. Review the rules (from the textbook) carefully. Knowing the rules makes everything simple.

3. Study the examples carefully and practice solving problems using both the math and the ray tracing technique. Do it for every possible arrangement of object and focal point (including) signs. It will all become very clear and simple and you will gain some intuition.

Hints for Solving Selected Problems

1-4. For plane mirrors remember the image distance is equal to the object distance. Reflected rays can be extended as straight lines to the other side of the mirror to form the virtual image.

5-11. Use the mirror formula. Magnification is given by $-q/p$. Convex mirrors have negative focal lengths.

12-24. Concave mirrors have positive focal lengths at $R/2$ on the side of R and are positive.

25-34. Use the single surface lens formula. For plane surfaces set $1/R$ equal to zero and solve for image distances using the formula.

35-59. Use the thin lens formula and remember that the magnification formula gives you a second equation relating p and q if M is known. When radii are given instead of focal length, use the lens maker formula. Draw sketches and use the rules for drawing the objects, lenses and images to guide you in obtaining an approximate solution and helping you to decide how to solve for the unknown.

24
Wave Optics

**Combining Light
Waves in Phase**

In this chapter we deal with wave optics which is the subject of how light interferes, diffracts, and can be polarized because of its wave characteristics.

24.1 Conditions for Interference

In order to achieve continuous interference of two or more light waves at least three conditions have to be met. They are:

1. The sources must be **coherent**; that is, they must **maintain a constant phase** with respect to each other. The best way of achieving this is to use a common, pinhole size source for all the beams or a laser.

2. The sources must be **monochromatic**, that is, of a single wavelength, so all the waves have the same length.

3. The superposition principle must apply (which it does for electric and magnetic fields).

24.2 Young's Double-Slit Experiment

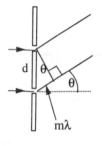

Thomas Young (1801) demonstrated the interference of a light wave passing through two slits that were close together. When the difference in distance traveled by the two beams, after emerging from the slits, was an integral number of wavelengths, positive or constructive interference was achieved. This means the waves added together and enhanced one another. When the distance was half of an odd integer of wavelengths the waves canceled and a dark spot on a screen placed some distance away was observed. For incident plane waves at $0°$, the difference in path is $d\sin\theta$, where d is the separation between the slits. Thus,

$$\text{path difference} = \delta = d\sin\theta = m\lambda \qquad \text{(24.1)}$$

where m = integer for constructive interference and m = (integer + 1/2) for destructive interference.

This can be expressed in terms of the geometry of the observing screen assuming small angles so that

$$\sin\theta \approx \tan\theta \approx \frac{y}{L} \qquad \text{(24.2)}$$

where y is the height on the screen and L is the distance to the screen. Bright and dark spots occur when

$$Y_{bright} = \frac{\lambda L}{d}m, \text{ and } y_{dark} = \frac{\lambda L}{d}\left(m + \frac{1}{2}\right), \text{ (m integer)} \qquad \text{(24.3)}$$

24.3 Change of Phase Due to Reflection

When light reflects from a surface of higher index of refraction, it goes through a phase change of 180° or π radians. This is equivalent to $\lambda/2$. If the index of refraction is less, then no phase change occurs. This is reminiscent of the reflection of a string hitting a boundary of a heavier string or lighter string respectively.

24.4 Interference in Thin Films

Interference can occur when a light ray (or wave) strikes a thin film and reflects off the front surface as well as the back surface and the two reflected waves are coherent but differ in phase because of the different path of each. To determine the condition for constructive interference we know the optical path difference must equal an integral number of wavelengths, and for destructive interference, the difference must be half a wave length in addition. Also, as described in the section above, any reflection at a boundary where the wave would enter a higher index of refraction, gives an additional half-wavelength optical path difference. It must also be remembered that in a medium with index of refraction n, the wavelength is $\lambda_n = \lambda_0/n$.

NEWTON'S RINGS

When a plano-convex lens is placed on top of a flat glass surface, one can get interference between the waves in the air-gap between the convex surface and the flat surface of the second piece of glass. The thin air wedge varies in thickness and thus a pattern of light and dark rings of differing color are produced. These interference fringes are called Newton's rings. The dark rings have radii given by $r \approx \sqrt{m\lambda\,(R/n)}$ where R is the radius of curvature of the upper lens and m is the number of the ring. Note: the center is the point of contact and is dark because the difference in distance in the air-gap is zero, but the reflections from the two surfaces still differ in phase by p, equivalent to half a wavelength.

24.5 Diffraction

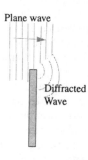

Plane wave

Diffracted Wave

Light waves passing edges bend around them in accordance with Huygens' predictions for waves. Similarly for passing through slits and holes and scattering around small objects, we observe the wave scattering around to the back. This effect is called *diffraction*. It can be regarded as interference from many coherent wave sources and that is what makes Huygens' principle work. If the slit is only a few times larger than the wavelength of the light, you will see an interference pattern of dark and bright bands on a nearby screen. If the screen is far enough away that the rays reaching the screen are approximately parallel, we call this Fraunhofer diffraction.

24.6 Single-Slit Diffraction

As light waves pass through a slit we can divide the width of the slit into an even integer of equal pieces, each of width $a/2n$. Then at an angle, θ, such that destructive interference occurs between corresponding points of adjacent pieces,

$$\frac{a}{2n}\sin\theta = \frac{\lambda}{2} \quad \text{for destructive interference}$$

$$\text{or simply} \quad \sin\theta = \frac{n\lambda}{a}$$

(24. 4)

Constructive interference maxima appear approximately midway between these minima, where $\sin\theta \approx n\lambda/2a$.

24.7 Polarization of Light Waves

For unpolarized light the electric field is oriented randomly but is transverse to the direction of propagation. For plane polarized light the electric vector is confined to a plane. There are several ways to polarize light.

POLARIZATION BY SELECTIVE ABSORPTION

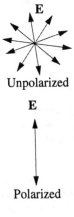

E

Unpolarized

E

Polarized

Some materials with long molecules lined up in a specific direction will only pass light whose vector direction is perpendicular with respect to the molecules. The others are absorbed. It is common to refer to the direction perpendicular to the molecular chains as the transmission axis. The light that comes through is said to be polarized and the material doing the polarizing is called the polarizer.

Once light passes through a polarizer and is polarized it will pass through a second polarizer without absorption if both polarizers are aligned and will be totally absorbed if they are oriented $90°$ with respect to each other. For an arbitrary orientation at angle θ, the intensity of the light transmitted is given by

$$I = I_0 \, (\cos\theta)^2 \qquad \text{(24.5)}$$

POLARIZATION BY REFLECTION

Reflected light is polarized to a certain extent if the angle of incidence is between 0° and 90°. If the angle of incidence is varied until it makes an angle of 90° with respect to the refracted beam, both the reflected and refracted beam will be polarized. The polarization of the E vector for the reflected beam is 100% and is parallel to the surface. The refracted beam is only partially polarized. This angle of incidence is called the polarizing angle θ_p or Brewster's angle and this relationship is known as **Brewster's law**. It can be written as

$$n = \frac{\sin\theta_p}{\sin(90^o - \theta_p)} = \tan\theta_p \qquad \text{(24.6)}$$

POLARIZATION BY SCATTERING

Scattering of light by the molecules of air can cause some polarization. Thus sunlight coming from different directions in the sky is partially polarized.

Short wavelengths (blue light) are more frequently scattered than long wavelengths (red light). This explains why clear sky appears blue and sunsets and sunrises appear red (the blue light has been scattered out). This effect is independent of polarization.

OPTICAL ACTIVITY

Some materials are optically active and rotate the plane of polarized light. Stressed plastic rotates the polarization more depending upon the amount of stress.

Some materials such as calcite crystals have two indices of refraction and each corresponds to a different polarization of the light. These crystals separate the orientations of polarization and give a double image and each image can be made to appear or disappear depending upon the orientation of the polarizer used to view the images.

24.8 Concept Statements and Questions

1. To interfere continuously with constant phase, two waves must be coherent (what does that mean?) and have the same wavelength (be monochromatic). Light from a point source or a laser passing through two slits or openings satisfy these conditions.

2. The light which emerges from two slits after satisfying 1. above, can constructively or destructively interfere depending on whether the difference in travel is an integer number of wavelengths or half-odd integer.

3. Under what condition will reflection of light undergo a phase change of 180°? Reflection from the two surfaces of a thin film can either constructively or destructively interfere depending upon the thickness of the film and the relative index of refraction compared to the two neighboring media.

4. Diffraction of light is an interference effect and is due to the wave nature of light.

5. In passing through a single slit, the waves coming from one zone of the slit will interact and interfere with waves from another zone. Thus, a one slit diffraction pattern is made.

5. Polarizing a light beam means to select from a beam all electric field vectors oriented in the same plane. Name four ways in which light can become polarized.

24.9 Hints for Solving the Problems

General Hints

1. Learn and thoroughly understand the formulae for interference from two slits. This physics is fundamental and is used in most of the problems in this chapter plus in later material.

2. Study all of the examples carefully.

3. It is important to take into account the optical path length difference, not the physical length difference. The optical path length takes into account the change in wavelength of the wave as it passes through the various media and also the change upon reflection when entering a more optically dense medium (higher index of refraction).

4. Study the problem-solving strategies for thin film interference given at the end of section 24.4.

Hints for Solving Selected Problems

1-16. Apply Young's double-slit interference formula. Also take into account the optical path difference as explained in General Hint 3. For small angles $\sin \theta \approx \tan \theta \approx \theta$(in radians). Draw sketches

17-37. These problems involve interference from reflecting thin films. To work these you have to remember the phase change of $\lambda/2$ or 180° when reflecting from one index of refraction to a higher one. You must also remember that the wavelength changes according to index of refraction. So it is the optical path difference that counts. Draw sketches of the setup. Assume (unless otherwise specified) normal incidence Also look at Examples and Problem-Solving Strategies in the text.

38-44. Draw sketches and then apply trig to solve for the angles and the components of the triangles. Use the equations giving the condition for destructive interference. Assume (unless otherwise specified) normal incidence

45-54. Use Brewster's law and Snell's law. Also you may need $I = I_0(\cos \theta)^2$.

25
Optical Instruments

**Making Use of
Light**

There is a great variety of optical instruments of very practical use. We discuss a few of the more common ones in this chapter.

25.1 The Camera

The single-lens camera consists of the lens, a light tight box, and film at the back to receive the image. By varying the distance between the lens and film the sharpest and most focused image can be found. A shutter allows for various exposure times making it possible to capture rapidly changing images.

The image size and brightness depends upon the focal length and aperture or lens diameter. A short focal length lens, for example, produces a small but bright image. The "light concentrating" power, therefore, depends on the f-number of the lens which is given by

$$f - \text{number} \ = \ \frac{f}{D} \qquad \text{(25. 1)}$$

A fast lens has a small f-number and vice versa for a slow lens. When the lighting is dim a fast lens is needed to gather enough light, so a large diameter lens is desirable.

25.2 The Eye

The eye is a very remarkable and complex organ and in many ways is similar to a camera. The front part of the eye is a curved membrane called the cornea where most of the light is refracted. This is followed by the aqueous humor and then a variable aperture (the iris and pupil) and finally the crystalline lens. The iris is colored and the f-number ranges from $f/2.8$ to $f/16$. This front system focuses the incoming light to a light sensitive membrane at the back of the eye called the retina. The retina picks up the image and transmits it to the brain along the optic nerve. Objects that are very close to the eye cannot be brought to a sharp focus. The nearest distance for which the image is still sharp is called the "near point."

DEFECTS OF THE EYE

Common maladies of the eye include farsightedness (hyperopia) when the image of a distant object is focused beyond the retina, nearsightedness (myopia) when the image is focused in front of the retina, and astigmatism, when the cornea and lens are not perfectly spherical and every point is focused as a line.

The eye is also subject to several diseases such as the formation of cataracts on the lens, which obscure and distort the image; glaucoma, which often damages the retina due to reduced blood supply because of increased pressure inside the eyeball; and detached retina, as well as inflammations due to bacteria and damage.

Lenses in prescribed corrective glasses are measured in diopters. The power, P, is the inverse of the focal length in meters, that is $P = 1/f$.

25.3 The Simple Magnifier

The simple magnifier is used to increase the apparent size of an object near the eye and it puts the image at infinity so the eye is relaxed for comfortable viewing. The angular magnification is ($m \equiv \theta/\theta_0$) where θ is the angle of the image and θ_0 ($\approx h/25$ cm) is the angular size of the object when at a person's near point (the distance from the eye and beyond where the image on the retina is clear, usually about 25 cm). If the lens of the magnifier is held near the eye and the object is brought mush closer to the eye than before, slightly closer than the far focal point, then the image is erect, magnified and at a large distance, thus comfortably viewed. The angle θ is now approximately h/f so the magnification is

$$m = \frac{\theta}{\theta_0} = \frac{h/f}{h/(25\text{ cm})} = \frac{25\text{ cm}}{f} \qquad \text{(25.2)}$$

25.4 The Compound Microscope

The compound microscope utilizes a short focal length objective lens and a simple magnifier for the eyepiece. The overall magnification is defined as the product of the lateral and angular magnifications where the lateral magnification is given as $M_l \approx -L/f_o$, so

$$M = M_1 m_e = -\frac{L}{f_o} \left(\frac{25 \text{ cm}}{f_e} \right) \qquad \text{(25. 3)}$$

The negative sign indicates that the image is inverted. The ultimate magnification of a compound microscope is limited by the wavelength of the light used to view the object and that is limited to the visible spectrum.

25.5 The Telescope

The refracting telescope uses a refracting lens both as the objective and the eyepiece. The angular magnification is given simply as

$$m = \frac{\theta}{\theta_0} = \frac{f_o}{f_e} \qquad \text{(25. 4)}$$

While the magnification can be increased with a short focal length eyepiece, all of the flaws, aberrations, and distortions are also magnified more, so the image may be blurred and of very poor quality. Quality magnification requires increasing the diameter of the objective lens to increase the resolution. In order to build very large objective lenses so that more light can be gathered and so a higher resolution can be obtained, it is convenient to build them as reflective, concave, parabolic mirrors. Thus many modern telescopes are of the reflecting type.

25.6 Resolution of Single-Slit and Circular Apertures

The wave nature of light limits the ability to distinguish between two closely spaced objects. One has to separate the diffraction pattern produced by light scattering from each. To decide whether there are two or only one object present we use the Rayleigh criterion. It can be stated as: *When the central maximum of one image falls on the first minimum of another image, the images are said to be just resolved.*

maximum

first minimum

Using the relationship for the first diffraction minimum, namely $sin\,\theta = \lambda/a$, and using the approximation for small angles that $\sin\theta \approx \theta$ we get the limiting angle of resolution of a slit to be

$$\theta_m = \frac{\lambda}{a}$$

(25. 5)

For a circular aperture with diameter D we get a slightly different answer, namely

$$\theta_m = 1.22\frac{\lambda}{D}$$

(25. 6)

25.7 The Michelson Interferometer

Invented by A. A. Michelson (1852-1931), the interferometer is an ingenious device that splits a light beam into two parts that travel in round-trip paths that are perpendicular to one another and then combines them upon return to form an interference pattern. It is used to make accurate length measurements. If one of the arms (light paths) is increased by $\lambda/4$, a constructive interference of the beam will change to destructive and vice versa. Twice this is called a fringe and by counting fringes it is easy to determine the increase in path length of one of the arms.

25.8 The Diffraction Grating

The diffraction grating is useful for producing line spectra (interference maxima for each wavelength present) of a light source. It consists of a large number of closely spaced slits (like 25,000 lines/inch).

The condition for interference maxima (for normal incidence) is

$$d\sin\theta = m\lambda \quad (m = \text{integer})$$ (25. 7)

RESOLVING POWER OF THE DIFFRACTION GRATING

The ability to separate two wavelengths that are nearly the same is called the resolving power. It is defined as $R \equiv \lambda/(\Delta\lambda)$ and is related to the number of lines, N, and the order of the image, m, as

$$R = Nm$$ (25. 8)

Note: For $m = 0$ all wavelengths are indistinguishable as this is the central maximum for all wavelengths.

25.9 Concept Statements and Questions

1. A small f-number for a lens implies that it lets in more light than if the f-number is large.

2. The eye is similar to a camera in many respects but is actually a very complex organ.

3. Farsightedness can be corrected with converging lenses of the right focal length. Similarly, nearsightedness can be corrected with the proper diverging lenses.

4. What is meant by the near point of the eye? How far away is it? How is it used in designing a simple magnifier?

5. How do the diameters of the objective lenses for compound microscopes and telescopes compare? Why do you suppose they are so different?

6. Is it possible to have a large magnification from a small telescope with a small diameter objective lens? What is wrong with such a telescope?

7. What does aperture size have to do with resolution in a magnifying system? Remember that the wavelength used is also important. So an electron microscope which uses very short wavelength matter waves can give much greater detail than an ordinary light microscope.

8. What is a good use of the Michelson interferometer?

9. How does the resolving power of a diffraction grating depend on the number of lines illuminated?

25.10 Hints for Solving the Problems

General Hints

1. Don't confuse focal length with f-number on a lens.

2. Be sure you understand the meaning of *resolving power* and for a circular opening how it depends on the diameter of the opening and the wavelength of the wave used.

3. The resolving power of a diffraction grating is for separating wavelengths close together, not images.

Hints for Solving Selected Problems

1-8. For a simple camera the ratio of image sizes is equal to the ratio of the object distance to image distance. The image distance is approximately the focal length. f-numbers are inversely proportional to the square root of the intensity. The time of exposure is inversely proportional to the intensity.

9-18. Single lens formula works for most of these. the power of the lens is the inverse of the focal length. The difficulty is in picking the image or object distances to satisfy the requested conditions.

19-24. Use the simple magnifier equation or simple lens formula.

25-36. Use the formulas for the compound microscope and telescope.

37-48. For circular openings the resolving power formula for Rayleigh's criterion is needed. To get the angular separation for some objects you can start with the lateral separation divided by the distance.

49-53. We get a fringe every time one of the arms of the interferometer changes length by half a wavelength.

54-64. Assume the light is of normal incidence, unless otherwise specified, and calculate the angles using the formula for constructive interference for various wavelengths. The resolving power is just Nm but it can be used to give $\Delta\lambda/\lambda$.

26
Relativity

**Physics at High
Speeds**

26.1 Introduction

Our intuition and expectations are based on our every
day experiences. At very high speeds the natural world
behaves somewhat differently than what we expect and
so we need to find out about these corrections. Because
motion is relative, the important predictions and consequences of high speed behavior is known as *"The Theory of Special Relativity"* or simply *"Relativity."* The
basic foundation for our understanding of relativity
was given to us by Einstein in 1905.

26.2 The Principle of Relativity

An inertial reference frame is one moving at constant
velocity. Newton's laws of motion are valid in inertial
reference frames. For everyday speeds the transforma-
tions of coordinates and speeds from one inertial reference frame to another are given so as to satisfy the

laws of mechanics. Thus there are no preferred reference frames.

26.3 The Speed of Light

Before the theory of relativity it was apparent that the laws of electricity and magnetism, in particular Maxwell's theory predicted the value of the speed of EM waves in the vacuum of free space, namely c. Yet if an observer of the EM wave had a velocity, v, that observer would expect to see his velocity added to the velocity of the wave in free space, hence he would no longer observe c but rather $c + v$. But then that makes some inertial frames different from others in violation of the principle of relativity (Maxwell's equations give c not $c + v$). Clearly something was inconsistent. Some thought there was a medium called *ether* through which the EM waves travel with a fixed speed and thus people began searching for the ether.

26.4 The Michelson-Morley Experiment

The Michelson-Morley experiment was performed in 1887 and was designed to show the relative motion of the earth with respect to the ether by studying the interference of a light beam with itself after being divided into two mutually perpendicular beams. One beam was directed parallel to the earth's motion about the sun and the other perpendicular to it. By rotating the apparatus through 90° the two arms were interchanged. If the speed of light varied with the Earth's motion, it would show up as a fringe shift in the interference beam. A null effect for the experiment was observed which indicated that the speed of light is unaffected by the motion of the observer and this seemed to rule out the idea of an absolute reference frame, equivalent to a theory needing an ether for the propagation of light.

26.5 Einstein's Principle of Relativity

Einstein made two postulates about nature and from those two postulates derived his theory of relativity and many consequences. The postulates and consequences are known as *The Special Theory of Relativity*. The two postulates are:

(1) The laws of physics take the same mathematical form in all inertial reference frames.

(2) The speed of light is independent of the motion of the source and the same in all inertial reference frames.

These postulates are in agreement with the null results of the Michelson-Morley experiment.

Accepting the statement "the speed of light is the same for all observers" is difficult because then displacements and time intervals will have to be different in one reference frame compared to another.

26.6 Consequences of Special Relativity

In relativity the events that occur are specified by three space coordinates and one time coordinate. The inclusion of time in specifying an event requires that clocks at all positions in a given inertial reference frame must be synchronized so as to read exactly the same time.

SIMULTANEITY AND THE RELATIVITY OF TIME

Two events that are separated spatially and are simultaneous in one reference frame may not be simultaneous in any other inertial reference frame that is moving with respect to the first.

TIME DILATION

According to an inertial observer who has a clock at rest with respect to him, if he sees another clock moving with respect to him, it runs slower even though identical in construction. This effect is known as time dilation and the comparative rates of the two clocks is given by

$$\Delta t = \frac{\Delta t'}{\sqrt{1 - v^2/c^2}} = \gamma \Delta t', \text{ where } \gamma = \frac{1}{\sqrt{1 - v^2/c^2}} \quad (26.1)$$

where $\Delta t'$ is measured on the "moving" clock, but at rest with respect to S', Δt is what the first observer measures the moving clock's time interval to be according to his stationary clock. "c" is the speed of light.

THE TWIN PARADOX

Since velocities and speeds are relative shouldn't two observers, say for example twins, in relative motion each see the other's clock going slower? How can they both be going slower? If one turns around and reverses his/her direction until he/she catches up to or returns to the starting point, what will they each conclude about each other's clocks? The twin who turns around and comes back undergoes accelerations and, therefore, will have less time lapse. That conclusion is reached by taking into account the accelerations of the one twin.

LENGTH CONTRACTION

Rocket at rest
(measured length)

Rocket at
high speed
(measured length)

Lengths measured by an observer in S' that are at rest in S' are denoted L' or L_p for "proper" length. When measured by an observer in S they are denoted as L and are found to be shortened by $1/\gamma$, i.e.

$$L = \frac{L_p}{\gamma} = L_p \sqrt{1 - \frac{v^2}{c^2}}$$ (26.2)

This contraction of the length occurs only along the direction of the motion. Lengths perpendicular to the direction of motion are unaffected.

26.7 Relativistic Momentum

The relativistic momentum to be a useful concept must (1) be conserved in collisions and (2) reduce to the nonrelativistic expression for momentum at low velocities, namely $m_0 v$. (m_0 is the mass of the particle when it's not moving.) The expression that satisfies these two conditions is

$$p = \gamma m_0 v \quad \text{where} \quad \gamma = \frac{1}{\sqrt{1 - (v/c)^2}}$$ (26.3)

26.8 Mass and the Ultimate Speed

The mass of an object increases with speed as observed from a reference frame at rest. The increase is given by

$$m = \gamma m_0 \qquad (26.4)$$

Using m as given the momentum is $p = mv$. As the velocity increases towards c, γ increases without bound. Thus c is the ultimate speed a particle can have.

26.9 Relativistic Addition of Velocities

Given the velocity u' along x in a reference frame moving with velocity v along x, the combined velocity as observed in the frame from which u' was measured is given by

$$u = \frac{u' + v}{1 + u'v/c^2} \qquad (26.5)$$

If either u' or v are along -x, we simply change their sign and as long as neither is greater than c to start with, then u also will be less than c. The denominator makes the difference between the nonrelativistic addition of velocities and the relativistic one.

26.10 Relativistic Energy

The total energy of a particle is found to be

$$E = \gamma m_0 c^2 = mc^2 \qquad (26.6)$$

So for $\gamma = 1$, $(u = 0)$, we get $E = m_0c^2$ (the energy of a particle at rest), Einstein's most famous equation. The kinetic energy of a particle is simply the difference between these two equations and is, therefore,

$$KE = E - E_0 = \gamma m_0 c^2 - m_0 c^2 = m_0 c^2 (\gamma - 1) \quad (26.7)$$

For low speeds compared to c, this expression reduces to the usual expression for kinetic energy, namely $KE = mu^2/2$.

Another useful expression for the total energy of a particle is

$$E^2 = p^2 c^2 + m_0^2 c^4 \quad (26.8)$$

This expression becomes $E = m_0c^2$ for $p = 0$, $(u = 0)$ just as before. It also gives $E = pc$ for $m = 0$, which is known to be true for electromagnetic waves, i.e., light.

26.11 Concept Statements and Questions

1. Relativity theory seems complex, but in many ways it is very simple. It is built on simple assumptions that have been verified experimentally and also seem very reasonable. Space and time are no longer absolutes and have been altered to make the postulates work. Study the postulates and question them until they are thoroughly understood. They are powerful statements about the world in which we live. They force us to accept the predictions of relativity theory. All the predictions have been verified to high accuracy.

2. Without $E = mc^2$, it would be difficult to understand nuclear energy, the formation of nuclei, fission and fusion, and how the stars (and our sun) live, evolve and eventually die. Particle accelerators and the experiments performed there would not make sense.

3. With c as the ultimate speed, we find that space, time, and mass cannot be absolute physical variables. With increasing speed, time dilates, lengths contract and mass increases according to a γ factor appropriately placed.

4. The relativistic increase in mass ($m = \gamma m_0$) shows up in the relativistic expression for momentum ($p = \gamma m_0 v = mv$) and total energy ($E_{tot} = \gamma m_0 c^2 = mc^2$.)

5. The addition of velocities is different because the ultimate speed cannot exceed c.

26.12 Hints for Solving the Problems

General Hints

1. When working with $\gamma = 1/\sqrt{1 - v^2/c^2}$ it is sometimes difficult to evaluate on your calculator. If $v << c$ then you can use the approximation $\gamma = 1 + v^2/2c^2$ or $\gamma^1 = 1 - v^2/2c^2$. The difficulty in some problems is to retain enough accuracy in the v^2/c^2 term to compare it with 1.0. Look for the possibility in these problems that a 1.0 will be subtracted off and thus eliminate your difficulty. Kinetic energy problems are like this and also some time dilation problems are like this.

2. Another hint is to recognize that $c^2 - v^2 = (c + v)(c - v)$ which is approximately $2c(c - v)$ for v close to c.

3. It may also be helpful in some evaluations of problems to notice that given a right triangle with the hypotenuse equal to c and one of the legs equal to v, the other leg is equal to γ^{1}. Thus sin θ, cos θ and tan θ relationships may be used.

Hints for Solving Selected Problems

1-20. Apply the basic time dilation and length contraction formulas given in Section 26.6 for each applicable part of the problem.

21-25. Use the formula for relativistic momentum in Section 26.7. Relativistic momentum is γ times the classical momentum.

26-32. To get γ take the total energy and divide by the rest mass energy. From γ one can get the velocity and vice versa. With γ you can figure out the length contraction and time dilation.

33-41. Use the formula for relativistic addition of velocities. Be careful in getting the sign of each of the velocities.

42-51. Use the same hints as for Problems 26-32.

27
Quantum Physics

**Physics on the
Atomic Scale**

The photon nature of light, which is strikingly different from its wavelike nature, has led to an understanding of nature that includes both particle and wavelike phenomena for all particles. The theory that describes what we know about this is called "Quantum Physics". It applies to all domains of physics but is usually not noticed except on the atomic or sub-atomic scales.

Only an introduction to the underlying ideas and principles of quantum physics can be given here since a thorough treatment takes us beyond the scope of this book.

27.1 Blackbody Radiation and Planck's Theory

Any object with a temperature, T, greater than zero Kelvin radiates away energy through the emission of electromagnetic radiation of all possible wavelengths.

It may also absorb radiation from the environment that happens to be incident upon it. Before the development of quantum theory, the classical approach to the theory was to explain the radiation as originating from accelerated charges due to small atomic sized oscillators and the thermal motion of the particles making up the object with the belief that the amount of energy emitted was continuous at all wavelengths. Several attempts to explain the observations in terms of conventional theory were unsuccessful, particularly the shape of the intensity versus wavelength curves. The theory was constructed for the theoretically perfect radiator or absorber called a "black body". One severe difficulty was the fact that there was a maximum in the curve, i.e., it turned over, and the location of the maximum was related to the temperature according to

$$\lambda_{max}T = 0.2898 \times 10^{-2} \text{ m} \cdot \text{K} \tag{27.1}$$

This formula is known as Wien's law.

A resolution to the dilemma occurred when Max Planck in 1900 speculated on the idea of quantized energy states of oscillators, which could transfer only *discrete* amounts of energy, namely

$$E_n = nhf \quad \text{where} \quad n = \text{integer} \tag{27.2}$$

and the energy of an emitted photon would correspond to the difference between two consecutive integer values of n so the photon would carry off an amount of energy given by

$$E = hf \tag{27.3}$$

This assumptions allowed Planck to fit the observed curves and to derive Wien's law from his formula by using calculus to locate the maximum.

Einstein later reinterpreted Planck's work concluding that the radiation itself was "lumpy", with energy given by $E = hf$. The assumptions of Planck and Einstein turned out to be useful for also explaining other peculiar atomic phenomena and they became the foundation of quantum theory.

27.2 The Photoelectric Effect

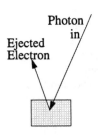

Photon in

Ejected Electron

Light hitting metals with enough energy will cause electrons to be ejected. The maximum energy of the ejected electrons is proportional to the frequency of the light and does not depend in any way on its intensity (energy in the light beam). Also a threshold frequency is observed below which no electrons are emitted regardless of the intensity. This effect, called the photoelectric effect, was successfully explained by Einstein using his theory of photons. His formula for the maximum kinetic energy of the electrons is given by

$$KE_{max} = eV_0 \quad \text{and}$$
$$KE_{max} = hf - \phi = hf - hf_{threshold}$$

(27.4)

V_0 is the voltage required to stop the most energetic electrons, so eV_0 corresponds to the energy of the most energetic electron and $\phi = hf_{threshold}$ is interpreted as the minimum energy required for an electron to break through and escape from the surface of the metal. ϕ is called the work function. The reason, according to the theory, that the kinetic energy of the electron is related to the frequency of the light and not the intensity is because the photons (a quantum of energy) carry $E = hf$ amount of energy, no more and no less. Increasing the intensity increases the number of photons, but not their individual energy. An electron when absorbing a photon, therefore, only absorbs hf amount of energy. For one electron to absorb more than one photon would be extremely unlikely.

The threshold wavelength can readily be obtained from $\lambda_{threshold} = c/f_{threshold}$.

27.3 Applications of the Photoelectric Effect

The photoelectric effect is the principle behind the operation of a photoelectric cell. The cell acts like a switch so when the light striking the cell is interrupted an electric circuit is turned on to open or close a door or perform some other useful task. The photoelectric cell is also used in the light meter of a camera to insure proper exposure and in a burglar alarm. It can also be used as a source of electricity where ample sunlight is available and perhaps someday will become one of the principal sources of our electrical power system.

27.4 X-Rays

X-rays when first discovered by Wilhelm Roentgen in 1895 were mysterious and their nature unknown. We now know that they are very energetic photons with wavelengths of about 0.1 nm. They can be diffracted from the atomic planes in crystals and produce interference patterns similar to ordinary light scattering from a diffraction grating.

X-rays are produced by the deceleration of energetic electrons and other charged particles. The maximum energy possible for the x-ray occurs when all of the energy of the electron is converted, i.e. $KE = eV = hf$ where V is voltage through which the electron was initially accelerated. The minimum wavelength would be

$$\lambda_{min} = \frac{c}{f_{max}} = \frac{hc}{eV}$$ (27.5)

Because of their high energy, x-rays are very penetrating but are also harmful to body tissue. They can be used to examine the inner structure of a living person, but the amount of dosage should be strictly monitored and controlled.

27.5 Diffraction of X-Rays by Crystals

For very short wavelengths such as x-rays we cannot make diffraction gratings with the line close enough together. It is useful instead to use the atomic planes of crystals and let the rays reflect off the various planes. The angles of diffraction for maxima are given by **Bragg's law** which is given by

$$2d\sin\theta = m\lambda \qquad \text{(m=integer)}$$ (27.6)

d is the spacing between the planes and θ is the angle of incidence which is equal to the angle of diffraction.

27.6 The Compton Effect

Another result similar to the photoelectric effect is to scatter photons from free or nearly free electrons such as valence electrons bound to the atoms of a gas. This effect is known as the Compton effect, named after Arthur Compton who discovered it in 1923. The observed effect is to witness that photons of energy $E = hf$ hit the electron and scatter off with energy $E' = hf'$ where $f' < f$. The energy of the electron is equal to the difference between E and E'. Energy is conserved, therefore, in the reaction and so is momentum. Thus the collision can be completely understood in terms of the kinematics of conserving energy and momentum. The momentum acquired by the electron is the vector difference, $\mathbf{p}' - \mathbf{p}$, between the initial momentum of the photon, $|\mathbf{p}| = E/c$ and the final momentum, $|\mathbf{p}'| = E'/c$. Expressing the results in terms of wavelength of the photons the equation obtained agrees very well with the observations and is

$$\lambda' - \lambda_0 = \frac{h}{mc}(1 - \cos\theta) \qquad \text{(27.7)}$$

where θ is the angle of the scattered photon.

27.7 Pair Production and Annihilation

The energy of a photon can, if high enough, be converted to a particle and anti-particle pair, for example, an electron and positron pair. The threshold energy is equal to the sum of the masses of the particles produced plus a small amount of energy that is given to a heavy particle which carries off the momentum for the threshold situation. Thus we get

$$E_{min} = hf_{min} = 2m_0c^2 \qquad \text{(27.8)}$$

Usually when a pair is produced the energy of the photon is above threshold so the particles carry off some kinetic energy in addition to their mass.

The anti-particle annihilates on a particle when it comes in contact. In the annihilation process two or more photons are given off. At least two are needed so that energy and momentum can both be conserved.

27.8 Photons and Electromagnetic Waves

At this point can we answer the question "Is light waves or particles?" The answer is, "both". Light is a string of electromagnetic waves that propagates best through a vacuum and comes in quantized amounts of energy and momentum. Thus, it acts as a particle when we're studying its mechanical properties involving scattering from single particles, but it interferes, refracts and diffracts like waves when interacting with holes or with multiple particle or macroscopic systems that are large compared to its wavelength. Neither characteristic depends upon intensity. A single photon will produce the same interference effects as an entire beam. The beam effect is really the sum of many single photons, each interfering with itself. The photon concept and the wave nature of light complement each other in giving us a more complete picture of what light is.

27.9 The Wave Properties of Particles

The dual nature of photons naturally leads to the question as to whether or not other particles have a wave like nature. Louis de Broglie wrote his doctoral dissertation on the topic and proposed the same relationship between wavelength and momentum for particles as existed for photons. That relationship is

$$\lambda = \frac{h}{p} = \frac{h}{mv} \qquad \text{(27.9)}$$

and is known as the de Broglie relationship. He also postulated a relationship between energy and frequency as $E = hf$, just like photons. **These matter waves are not electromagnetic in nature.**

27.10 The Wave Function

Erwin Schrödinger gave us a way to determine the behavior of the waves associated with particles. His wave equation has been applied successfully to several quantum systems including the hydrogen atom. His theory is even more successful than Bohr's theory of hydrogen which was limited to selected properties of hydrogen.

Electrons scattering through a double-slit arrangement can be sent one at a time, and the electrons when hitting the screen each make a bright spot at only one location. After sending many electrons one at a time through the slits a pattern of where the electrons arrive eventually builds up to produce maxima and minima in exact accordance with the photon equivalent double-slit experiment of the same wavelength, namely

$$d \sin\theta = m\lambda \quad \text{for maxima,} \quad (\text{m = integer}) \qquad \text{(27. 10)}$$

The dual nature of the electron is clearly shown. The electrons detected on the screen are detected as particles but their distribution is governed by the interference of waves.

If one slit is covered, a single-slit diffraction pattern is produced for each of the slits left uncovered. The sum of the two single-slit patterns will not give the two-slit pattern, because there is no interference. To get the interference, the wave associated with the electron must pass through both slits simultaneously. That means the electron passes through both slits, which is a difficult concept to comprehend for a particle but not for a wave.

27.11 The Uncertainty Principle

Because of the probability wave associated with particles in the quantum theory, it became clear that the position of the particle is uncertain, (Δx), to about the size of the wavelength, and since the momentum, according to de Broglie, is the inverse of the wavelength, the momentum would also be uncertain, (Δp_x). Werner Heisenberg in 1927 showed that the product of these two uncertainties is always as large or larger than $h/4\pi$. The Heisenberg uncertainty principle thus states

$$\Delta x \Delta p_x \geq \frac{1}{2}(h/2\pi) \qquad \text{(27. 11)}$$

(This principle can also be applied to the uncertainty in energy and uncertainty in the time of a system, i.e., $\Delta E \Delta t \geq h/4\pi$ and to other variables in a quantum system that are related in a certain similar way .)

27.12 Concept Statements and Questions

1. The idea of quantized energy transfer was introduced by Planck to explain black-body radiation. The energy of a photon is proportional to its frequency and hence its momentum is inversely proportional to its wavelength.

2. Einstein introduced the idea of photons to explain the photoelectric effect. He also predicted the threshold frequency as a consequence of the work function of the metal.

3. The Compton effect of photon scattering from "free" electrons further confirms the idea of "quanta of energy".

4. De Broglie introduced the theory that all particles in nature have a wave-like phenomenon associated with them with the wavelength being related to the momentum and the frequency related to the energy being the same as with photons.

5. Particles can be diffracted by a crystal lattice and produce interference patterns just like x-rays. Is this evidence that particles have a wave-like characteristic?

6. If a photon has enough energy it can produce a particle anti-particle pair. How much energy would be required to produce a proton anti-proton pair? ($m_p c^2 = 938.6$ MeV.)

7. Heisenberg used the wave nature of particles to give us the uncertainty principle which places fundamental limitations on the precision with which we can know a particle's position and momentum simultaneously.

8. A double-slit interference experiment using only one electron at a time gives interference patterns showing that the electron wave travels through both slits.

9. The Schrödinger equation gives a complex-valued wave function as a solution whose absolute square gives the probability of finding a particle at a position x.

10. Tunneling through barriers is another important prediction of quantum theory and it is observed in various situations.

27.13 Hints for Solving the Problems

General Hints

1. The work function for the photoelectric effect is usually given in electron volts, so it is most convenient to choose h (Planck's constant) in terms of these units.

2. The threshold energy for the photon in the photoelectric effect is the work function.

3. When working with the uncertainty principle in a problem, the region to which an object is confined (such as the radius of the nucleus for a nuclear system) can be considered as the uncertainty of its position. Similarly, for momentum the range of possible values, as restricted by the system, becomes the uncertainty.

Hints for Solving Selected Problems

1-10. Use Wien's law from Section 27.1. You can determine the most prominent type of radiation from the peak wavelength. The number of photons associated with a given amount of radiant energy can be determined by finding the energy associated with each photon. To solve for n, therefore, use $E = nhf$ and consider E and f as givens.

9-18. Apply Einstein's photoelectric effect formula. Use h in units of eV· s. then you don't have to convert. Remember, the threshold energy to remove an electron is the work function.

21-24. The energy of x-rays can be obtained from either their frequency or wavelength. From $E = qV$ where $E = hf$, you can relate frequency to potential difference.

25-30. Convert electron energies and momenta to wavelengths then use the double-slit formula just like you did for light.

31-38. Use Compton's formula.

39-43. In annihilation processes the photons produced carry off all the energy and momentum available in the center of mass. If two photons are produced, therefore, each takes half the total available energy as seen in the center of mass system.

26-36. Use the de Broglie formula for changing from momentum to wavelength and vice versa. Once you have the momentum you can get the kinetic energy of a particle and vice versa.

55-61. Uncertainty in measurement of x or p can be used in the uncertainty principle to put limits on the other variable such as velocity, energy, and time.

28
Atomic Physics

The Quantum Theory Applied to Hydrogen

Applying the complete theory of quantum mechanics to the hydrogen atom gives a great deal more information and insight than was possible with Bohr's simple model. In this chapter we will gain understanding as to the physical meaning of the various quantum numbers and how the theory applies to the more complex atomic systems. We will also look at the *Exclusion Principle* given to us by Wolfgang Pauli.

28.1 Early Models of the Atom

Before Newton the atom was considered to be a hard impenetrable sphere or objects of other shapes. When the electrical nature of the nucleus (the part of the atom that had the positive charge) and electrons was discovered, J. J. Thompson devised a model of the atom which consisted of a volume of positive charge with negative electrons scattered throughout. Then Ernest Rutherford showed that the positive charge was associated with a massive (compared to electrons), extremely dense, small nucleus and with the electrons somewhere outside.

Using Rutherford's ideas Bohr went ahead and devised a model of the atom that successfully explained Rutherford's observations and also properly accounted for the spectral lines coming from hydrogen and similarly arranged nuclei. Finally, quantum mechanics was applied to the hydrogen atom along with relativistic corrections and these theories have been the most successful. They account for many additional observed data including subshells and hyperfine structure in the spectra.

28.2 Atomic Spectra

When a tube filled with a low-pressure gas of some element is energetically excited in an electrical discharge it glows and if that light is viewed through a diffraction grating, a set of brightly colored lines of varying colors is observed. Each element has its own set of lines that characterize that element. In 1885 Johann Balmer discovered a formula for the wavelengths of certain spectral lines observed coming from excited hydrogen. The formula is:

$$\frac{1}{\lambda} = R_H \left(\frac{1}{2^2} - \frac{1}{n^2} \right) \quad \text{for} \quad n = 3, 4, 5, \ldots \qquad \text{(28.1)}$$

and $R_H = 1.0973732 \times 10^7$ m^{-1} is the Rydberg constant. After Balmer's success, some others found similar formula's for other sets of hydrogen lines. These include the Lyman, Paschen, and Brackett series.

When a continuous spectrum of white light is passed through a cool sample of gas of an element (such as hydrogen), a set of dark lines shows up instead of the bright lines seen for the hot gas. These dark lines are due to absorption of certain wavelengths.

28.3 The Bohr Theory of Hydrogen

Niels Bohr developed a theory of quantized orbits and energy levels for electrons in the hydrogen atom which was able to match the observed spectral lines coming from the element and also predict others. His assumptions were:

1. The electrons move in circular orbits and the centripetal force is provided by the Coulomb force of attraction between the proton and electron.

2. Only certain orbits are allowed and the condition on the orbits is that the angular momentum can only take on integral units of the fundamental unit of angular momentum, namely $L = nh/2\pi$ where $n = 1$, 2..., a positive integer.

3. Energy is emitted or absorbed when the electron jumps from one level to another. for a transition from a higher orbit to a lower orbit the photon is emitted and the energy difference is hf where f is the frequency of the photon. Photon absorption causes a transition of the electron to a higher orbit but only if the energy and frequency are just right.

The total energy of the electron is

$$E = KE + PE = \frac{1}{2}mv^2 - k\frac{e^2}{r} = -k\frac{e^2}{2r} \qquad \text{(28.2)}$$

The application of orbits of quantized angular momentum, $mvr = nh/2\pi$, gives the following condition on their radius

$$r_n = \frac{n^2 h^2}{4\pi^2 m k_e^2} \quad \text{where} \quad n = 1, 2, 3 \ldots \quad \text{(28.3)}$$

Thus the total energy could be obtained in terms of the quantized orbits by substituting in r_n.

$$E_n = -k_e \frac{e^2}{2a_0} \left(\frac{1}{n^2}\right) = -\frac{13.6}{n^2} \text{ eV} \quad n = 1, 2, 3, \ldots \quad \text{(28.4)}$$

where a_0 is the value of r_n when n = 1 and equals the radius of the lowest possible orbit, namely, 0.0529 nm.

The frequency of an emitted photon corresponds to the energy difference between two levels, so

$$f = \frac{E_i - E_f}{h} = \frac{ke^2}{2a_0 h} \left(\frac{1}{n_f^2} - \frac{1}{n_i^2}\right) \quad \text{(28.5)}$$

An equation for the wavelength, λ, can be obtained from this since $f\lambda = c$. It is

$$\frac{1}{\lambda} = R_H \left(\frac{1}{n_f^2} - \frac{1}{n_i^2}\right) = 1.0973732 \left(\frac{1}{n_f^2} - \frac{1}{n_i^2}\right) \times 10^7 \quad \text{(28.6)}$$

where the Rydbery constant R_H has the experimental value as given in units of meter^{-1} but also is given entirely by $ke^2/2a_0 hc$. Thus Bohr was able to derive Balmer's formula and also the others for hydrogen that were similar in form.

[diagram of energy levels:]

0

-13.6/n^2

-1.51 eV

-3.4 eV

-13.6 eV

Energy levels in hydrogrogen

The theory applies equally well to any other atom with only one electron such as ionized helium or doubly ionized lithium.

BOHR'S CORRESPONDENCE PRINCIPLE

Quantum mechanics is in agreement with classical physics where the energy differences between quantized levels becomes vanishingly small.

28.4 Modification of the Bohr Theory

Even though Bohr's theory of the atom was very successful, there were certain important discrepancies that could not be explained. Sommerfeld helped by extending the theory to include elliptical orbits. He also included a new quantum number, l, an integer, to account for various subshells (distinguished by angular momentum) where l ranges from 0 to n-1 and specifies the relative amount of angular momentum, and n is the principal quantum number designating the shell. Later it was found that a magnetic field would split the spectral lines corresponding to a given subshell l, so a magnetic quantum number m_l was introduced which ranges from $-l$ to $+l$ in integer steps. Finally, additional fine splitting was observed which could be attributed to the "spin" of the electron which could have two orientations, up and down, for each magnetic quantum number, m_l.

28.5 De Broglie Wave and the Hydrogen Atom

De Broglie's equation, $\lambda = h/p$, neatly explains why only certain orbits and energy levels are allowed in atoms according to Bohr's model. That is because the allowed orbits are those which set up standing waves for the electrons, i.e., an integral number of de Broglie waves will fit in the circumference of the electron's orbit so that $n\lambda = 2\pi r$ is satisfied. This condition gives Bohr's quantization rule that $L = mvr = nh/2\pi$. This interpretation is interesting but was soon shown to be unsatisfactory. For example, it doesn't allow for various angular momentum states associated with the l quantum number.

28.6 Quantum Mechanics and the Hydrogen Atom

Applying the theory of quantum mechanics to the hydrogen atom gives a much more detailed and satisfactory description than Bohr's theory. The solution of the Schrödinger wave equation is three dimensional and agrees with the successes of the Bohr theory but it also goes much beyond and gives two more quantum numbers in addition to n. n is called the principal quantum number and specifies the shell that the electron is in. Then we have l and m_l which are called the orbital and the orbital magnetic quantum numbers, respectively. n can range from 1 to ∞; l can range from 0 to $n-l$; and m_l can range from $-l$ to l. The shells are named with capital letters of the alphabet and begin with K for $n = 1$ and go on with L, M, N,. . . for higher n. All electrons with the same values of n and l are designated as subshells using the letters $s, p, d, f, g, h \ldots$ for which $l = 0, 1, 2, 3, \ldots$ States are specified with the number n followed by the letter for the subshell. For example, $5d$ means $n = 5$ and $l = 2$.

28.7 The Spin Magnetic Quantum Number

It has been discovered that electrons have an intrinsic quantum number called spin and designated with m_s. The values of spin are $+1/2$ and $-1/2$ and have the same units as l and m_l which are the same as $h/2\pi$. With the inclusion of spin, the number of possible states each with a different set of quantum numbers in an atom is doubled. There is one set for positive spin and another set for negative spin.

28.8 Electron Clouds

According to quantum mechanics the solution to the wave equation, ψ, is not an equation of motion for the particle, but rather an equation describing the behavior of the particle's wave. As such, since the wave is "smeared" over a region of space, the location is not well defined until a measurement is made. The wavefunction squared gives the probability of finding the electron within a small distance, Δx, of the chosen position, x. Locating the electron at x also gives the values of the quantum numbers associated with that particular position but certain other quantum numbers are incompatible with the measurement, for example, the momentum for which there is an uncertainty, Δp.

28.9 The Exclusion Principle and the Periodic Table

The quantum numbers associated with each electron include n, l, m_l, and m_s, (the principal, the orbital angular momentum, the magnetic and the spin quantum numbers). How many electrons in an atom can have the same set of quantum numbers? Wolfgang Pauli answered that question and said "only one." This statement is known as the *Pauli exclusion principle*. This principle insures that every electron has its own domain or kingdom and since particles always tend to go to the lowest energy state allowed, as more electrons are added to an atomic system, all of the lowest energy shells and subshells fill up. In following the rules for the allowed values of the quantum numbers, it is found that the electron configurations in atoms match well with observation and this enables us to understand the periodic table of the elements arranged according to chemical properties for us by Dmitri Mendeleev in 1871. Thus much of the chemistry and physics of each of the various elements is not only understood but predictable. The electronic configuration can be symbolically represented and the way the shells fill up determines the chemical properties of the elements. As examples, the electronic configuration of beryllium with its four electrons is $1s^2 2s^2$, and boron has a configuration of $1s^2 2s^2 2p^1$. The superscripts refer to the number of electrons in that subshell.

Atoms with filled shells tend to be chemically inert. (See the text for more details on the electronic configurations and how they are determined.)

28.10 Characteristic X-Rays

We explained earlier the origin of x-rays as being associated with accelerating electrons or other charged particles. There are also spectral lines originating in atoms with energies of x-rays and these are called "characteristic x-rays." They arise in elements of high atomic number where the innermost electrons make transitions between the inner shells for which the energies are highest in the atom. If the lowest shell (K shell) loses an electron because of collision or for any other reason, then electrons from higher shells (L shell ($n=2$), M shell ($n=3$), etc.) make a transition to fill the K shell vacancy and thus emit a photon which has x-ray energy denoted K_α, K_β x-rays etc. respectively. Now a new vacancy occurs in the L or M or N shell etc. which can be filled by electrons cascading down from even higher shells and so on until the atom is back to normal, i.e. all the electrons are in the lowest quantum states possible.

For the K shell there are normally two electrons. If one of the electrons is missing, an electron from one of the outer shells sees an effective charge of Z-1 due to shielding from the remaining electron. The energy of the K shell would now be approximately

$$E_K = -(Z-1)^2 13.6 \text{ eV} \qquad \text{(28. 7)}$$

Henry G. J. Moseley in 1914 was able to use ideas related to this formula to figure out the Z value of unknown elements and thus help complete the chart of the elements.

28.11 Atomic Transitions

The electrons in an atom will absorb a photon only if the photon energy corresponds to an energy differences between where the electron starts out and where it ends up, i.e., E_f - E_i.

After a short time (~10^{-8} s) the electron makes a transition to any lower state that is not filled and that is also consistent with the transition rules. If another photon comes by that has exactly the energy of the transition, it will *stimulate* the electron to make the jump. Both the photons will go off together parallel and in phase. It is this process that makes the laser possible.

28.12 Lasers and Holography

The word laser is an acronym for *l*ight *a*mplification by *s*timulated *e*mission of *r*adiation and requires a mechanism (energy input from an oscillator as part of the laser, for example, causing electron collisions with the atoms of the gas and exciting them) for putting an over-population of electrons into a higher energy *metastable state* than would occur naturally from thermal equilibrium. Incoming photons with the right frequency will then stimulate these excited electrons. The more photons released, the more photons there are to do even more stimulating. So a chain reaction occurs and thus the light is amplified and is also coherent and collimated. To enhance the number of photons to do the stimulating, reflecting mirrors are put on the ends of the system with one being slightly transparent to allow a fraction of the photons to escape. Multiple reflections builds the photon count up to a high intensity.

HOLOGRAPHY

Lasers are used to make holograms which are photographic recordings of the interference of two coherent beams, one reflecting from the object and the other taking another path. Illuminating the hologram with the same frequency of light causes a three-dimensional image of the original object to appear.

28.13 Concept Statements and Questions

1. State the assumptions in the Bohr theory of the atom. How well did his theory explain the facts known about the atomic spectra of hydrogen in his day?

2. Four numbers (n, l, m_l, m_s) are needed to specify the state of an electron in an atomic system. What does each of the numbers represent?

3. How do de Broglie waves fit into the quantum conditions Bohr required for angular momentum?

4. Why do we speak of electron clouds when talking about the positions of electrons in atoms?

5. State the Pauli exclusion principle in your own words and explain how it operates in atoms.

6. The electron spin is an intrinsic angular momentum associated with all electrons. Its projection along a specified axis is quantized with values of $m_s = \pm 1/2$ (h/

2π). Due to this spin an electron also has a magnetic moment.

7. X-rays can be produced by transitions to the low-lying states in an atom of high atomic number (so the transition energy is high enough to be called x-ray) when there is a vacancy.

8. Electrons can be <u>spontaneous</u> in their transition from a higher to a lower state, or they can be <u>stimulated</u> by a photon of the same frequency to make the transition.

9. *Laser* is an acronym. What does it stand for and how does a laser work as explained in simple words?

28.14 Hints for Solving the Problems

General Hints

1. Learn the Bohr model of the atom and the postulates accompanying the model. It is also very useful to memorize Equation 28.4. Study Examples 28.1 and 28.2 in the text.

2. Be sure you know, according to quantum mechancis applied to the atom, what values are possible for each of the quantum numbers.

3. Learn the classification scheme for the shells and subshells and their alphabetic letter representation.

4. Be sure you understand how an electronic transition gives rise to the emission or absorption of a photon. The photon carries energy equal to the difference in energies between the two states.

Hints for Solving Selected Problems

1-7. Use the Balmer formula for computing wavelengths of the photons and then frequencies and energies. Remember the electrostatic force formula. Also remember the total energy formula of an electron in orbit about the proton and how it is made up of the kinetic energy plus potential energy terms.

8-27. Use Bohr's formula for the energy of the ground state of the hydgrogen atom. The energy of the excited states is given as $E_0/n^2 = -13.6/n^2$. See General Hint 4.

28-34. Apply de Broglie's ideas about the circumference of an electron orbit being equal to $n\lambda$ to explain Bohr's angular momentum quantization rule. Also atoms that have been stripped of all their electrons except one are hydrogen-like so Bohr's formula applies providing Z as the increased charge on the nucleus is taken into account.

35-41. In order to count states you need to know the limits on the quantum numbers. Every possible combination of numbers allowed within the limits is another state. Be sure to include the spin of the electron if all states are to be counted. In the case of spin = 1 like the ρ-meson, there are three oreintations of the spin instead of just two.

42-45. The characteristic x-rays are just unusually energetic photons due to electronic transitions.

29
Nuclear Physics

**Deep, Deep
Inside the Atom**

The atomic nucleus is exceedingly small compared to the size of the atom, 10^{-4} to 10^{-5} times smaller. Yet that is where nearly all of the mass is found. Radioactivity, discovered by Antoine Henri Becquerel in 1896, and other instabilities of the atom are found to come from the nucleus. Rutherford identified three types of radiation called alpha, beta, and gamma (the first three alphabetic letters in greek) coming from the nucleus.

In this chapter we'll discuss some of the properties of nuclei, what we know about their composition and structure, about nuclear reactions and how they disintegrate or decay.

29.1 Some Properties of Nuclei

Two types of particles, protons and neutrons, are found in nuclei. The number of protons is what we have called atomic number, Z. The number of neutrons is N.

The sum of Z plus N is called the atomic mass number, A. The symbols used to represent nuclei are the chemical symbols identifying the element. Nuclei with a fixed number of protons but which vary in the number of neutrons are called *isotopes*. A complete symbolic representation includes as superscripts and subscripts on the chemical symbol the atomic mass number and atomic number.

CHARGE AND MASS

Each proton carries 1 electronic unit of positive charge. Masses are frequently given in energy equivalent units, namely $E = mc^2$. So instead of giving m, E is given instead. Thus the mass of a proton is $m = 1.67 \times 10^{-27}$ kg or 938.2 MeV. In energy units the mass of the electron is 0.511 MeV and the neutron is 939.6 MeV.

The unified mass unit (1/12 of the mass of carbon 12) is $1\ u = 1.220559 \times 10^{-27}$ kg = 931.50 MeV.

THE SIZE OF NUCLEI

Scattering experiments have enabled us to determine that the proton has a radius, r_0, of about 1.2 fm = 1.2×10^{-15} m. For larger nuclei the radius is approximately

$$r = r_0 A^{1/3}$$ (29.1)

The volume is proportional to A, showing that all nuclei have the same density so there is little or no compression as we increase the mass.

NUCLEAR STABILITY

The Coulomb repulsion between protons opposes the nuclear binding of the strong force. Therefore, only certain combinations of nuclei are stable. There are about 400 stable nuclei. For light nuclei the stable ones have N approximately equal Z, whereas as you go up to higher and higher A, the number of neutrons become more and more numerous relative to Z.

For certain numbers of N and Z, namely 2, 8, 20, 28, 50, 82, and 126, called magic numbers, the nuclei are most stable.

Nuclear Spin and Magnetic Moment

The intrinsic spin of a nucleus is made up of orbital angular momentum of the nucleons (neutrons and protons), which are integers, added to the intrinsic spins of the nucleons which is exactly the same as an electron. Thus, the intrinsic spin is either an integer or half an odd integer.

The magnetic moment of a nucleus is measured in terms of the nuclear magneton, μ_n, defined as

$$\mu_n \equiv \frac{e}{2m_p}\left(\frac{h}{2\pi}\right) = 5.05 \times 10^{-27} \text{ J/T} \qquad (29.2)$$

The magnetic moments of the proton and neutron in units of the nuclear magneton are 2.7928 and -1.9135, respectively. These unexpected values are believed to be connected with the internal structure of the nucleons.

Because of their magnetic moments nuclei will precess in a magnetic field. The energy of the state is dependent upon the orientation. Transitions between these various states form the basis for nuclear magnetic resonance or NMR. This phenomenon can be used in medicine to get an image, nuclear magnetic imaging or NMI, of an internal organ. If hydrogen is used, we get the location of concentrations of hydrogen.

29.2 Binding Energy

Due to binding energy which is negative, a nucleus made of several nucleons will have less mass than the sum of the masses of the individual nucleons. The difference in masses, Δm, times c^2 gives the binding energy in energy units.

Because of the binding energy, light elements when fused together will release energy in the fusing process. Elements heavier than iron have less and less average binding (binding energy per nucleon) so fusion requires energy to make it happen. The opposite process of fission, however, releases energy for the most massive nuclei.

29.3 Radioactivity

Radioactivity is a process of spontaneous emission of particles or radiation from the nucleus. The most common types of radiation are alpha, beta and gamma. The alpha is a helium nucleus of 2 protons and 2 neutrons. The beta particles are either electrons or positrons (antielectrons with a positive charge) and the gammas are high energy photons which come from an excited state of the nucleus. Clearly the emission of alpha particles or beta rays changes the charge of the nucleus and the emission of alphas significantly changes the mass.

The rate at which a particular decay process occurs is proportional to the amount of material or the number of nuclei, N, that you start with, so the number that decay, ΔN in the time Δt is given by

$$\Delta N = -\lambda N \Delta t \qquad \text{(29. 3)}$$

where λ is the decay constant and the minus sign indicates that the number of nuclei remaining is decreasing. Solving this equation by integration is simple and gives

$$N = N_0 e^{-\lambda t} \qquad \text{(29. 4)}$$

The decay rate, $R = |\Delta N/\Delta t|$ is simply λN, but N is given in Equation 29.4 so

$$R = \lambda N = \lambda N_0 e^{-\lambda t} = R_0 e^{-\lambda t} \qquad \text{(29. 5)}$$

The half-life is the time it takes 1/2 of the material to decay and it is related to λ by

$$T_{1/2} = \frac{\ln \lambda}{2} = \frac{0.693}{\lambda} \qquad \text{(29. 6)}$$

Expressed in terms of the number of half-lifes, n, that have transpired, the decay law becomes $N = N_0 2^{-n}$.

The unit of radioactivity is the curie (Ci), defined as 1 Ci = 3.7 x 10^{10} decay/s. The SI unit of activity is called the becquerel (Bq) and it is, 1 Bq = 1 decay/s.

29.4 The Decay Processes

ALPHA DECAY

If a nucleus emits an alpha particle it loses two protons and two neutrons. This can be represented symbolically by

$$^A_Z X = \ ^{A-4}_{Z-2}Y + \ ^4_2He \tag{29.7}$$

The disintegration energy, Q, is defined as

$$Q \equiv (M_X - (M_Y + M_\alpha))\,c^2 \tag{29.8}$$

Q could be worked out in terms of atomic mass units and then converted to energy by $1\ u = 931.50$ MeV.

The mechanism of the alpha escape is tunneling through the Coulomb barrier.

BETA DECAY (Emitting an electron or positron)

The emission of a β^- increases the atomic number by one and the emission of a β^+ decreases it by one. Also when a β particle is emitted, a neutrino (or antineutrino for β^-) is emitted. The neutrino has negligibly small rest mass and has no charge but does carry away energy, momentum and spin angular momentum. A typical reaction involving both neutrinos and beta particles is the decay of the free neutron given by

$$n \rightarrow p + \beta^- \ + \bar{\nu} \tag{29.9}$$

GAMMA DECAY

Gamma decay is the emission of a very energetic photon. This kind of decay only occurs if the nucleus is in a very excited state, like what might occur if some other decay has already taken place.

If the energy of the nucleus is excited sufficiently high it may emit neutrons or protons or other combinations of these particles in order to get to a stable state.

ELECTRON CAPTURE

A process that competes with beta decay is electron capture from the K or L shells of the atom. A neutrino is given off.

PRACTICAL USES OF RADIOACTIVITY

Carbon Dating. The β^- decay of ^{14}C with a half-life of 5730 years is used to date organic materials. The radioactive carbon is picked up from the atmosphere by living organisms and when the organism dies, it gradually decays away. It is fairly reliable as a dating procedure up to 25,000 years.

Smoke Detectors. Some smoke detectors use an ionization chamber and a small radioactive source to provide ionization in the chamber allowing a trickle current to flow until it is interrupted by large smoke particles, at which time an alarm sounds.

29.5 Natural Radioactivity

Three series of radioactive nuclei occur naturally. They begin with ^{238}U, ^{235}U and ^{232}Th and are known as the uranium, actinium, and thorium series. A fourth series starting with artificially produced neptunium (^{237}Np) is also known.

There are also several other radioactive isotopes found in nature besides those in the series.

29.6 Nuclear Reactions

Nuclear particles that collide with other nuclei can produce all kinds of reactions. The general type reaction is

$$a + X = Y + b \qquad \text{(29. 10)}$$

The total energy released (exothermic) or absorbed (endothermic) by the reaction if all particles are initially at rest is given as the reaction energy, Q, as

$$Q = (M_a + M_X - M_Y - M_b)\, c^2 \qquad \text{(29. 11)}$$

An endothermic reaction occurs only if the bombarding kinetic energy is somewhat greater than Q. The lowest energy at which this is possible is called the threshold energy.

If the particles a and b are identical, the reaction is called scattering.

29.7 Concept Statements and Questions

1. Nuclei are extremely small compared to the atoms and are made of protons and neutrons.

2. The density of all nuclei is the same and hence the radius is proportional to the atomic number, A.

3. Nuclei have an intrinsic angular momentum called nuclear spin. Both integer and half-odd integer values for spin are possible. Zero is very common.

4. The binding energy of nuclei is so great that there is a measurable difference in mass between the nucleus and the sum of its constituents when not bound.

5. The decay rate of radioactive material is proportional to the number of radioactive atoms present.

6. The particles emitted in radioactive decay are alpha particles, beta particles, or gamma rays unless the nucleus is highly excited, whereupon the emission of other particles is possible.

7. The Q of a reaction tells how much energy is involved in the reaction.

29.8 Hints for Solving the Problems

General Hints

1. Simplify calculations for nuclear particles by using the atomic mass unit and then convert over to energy using the conversion for $u = 931.5$ MeV/c^2.

2. Binding energy is equal to the mass deficit converted to energy units. The mass deficit is the difference between the mass of the particles unbound and when they are bound..

3. One key to figuring out nuclear reactions is to keep track of all the things that have to be conserved such as charge, nucleon number, angular momentum, energy

etc. Most of the conservation laws are simply additive and deal with small numbers.

Hints for Solving Selected Problems

1-13. The radius of nuclei can be calculated from the radius formula. The density is a constant so once you figure it for one nucleus it will be the same for all. If comparisons are made take a ratio so the density factor drops out. Remember that density is mass/volume

14-22. The binding energy per nucleon is the total binding energy divided by the number of nucleons. Use the binding energy formula and the number of nucleons. For all the problems involving binding energy you need to look up the atomic masses in tables in the text.

23-35. Use the radioactive decay formula. The activity is given by λN.

36-45. Keep track of the charges and number of nucleons to balance the equations.

46-49. Count charges and numbers of nucleons to figure out unknown reactants. Conserve energy and use Equations 29.10 and 29.11 for the formulas for calculating Q.

30
Energy and Elementary Particles

Making Physics at the Smallest Useful

We have considered atoms and nuclei, but now we look at fission and fusion, two techniques for extracting nuclear energy. Fusion puts two lighter nuclei together to make a heavier one and works well at releasing energy up to iron. After iron, fission, which breaks a large nucleus into two or more smaller ones, releases energy. It is also of great interest to study the structure of the basic constituents of the elementary particles as we consider the theories and properties of quarks and leptons.

Oddly, an understanding of the fundamental particles and accompanying forces and interactions gives us tremendous insights into the universe and how it began and how it evolved to its present state.

30.1 Nuclear Fission

Extracting energy from nuclear fission processes requires that a heavy nucleus like uranium be broken into smaller nuclei and fragments that are, on the average, more tightly bound. Thus the total rest mass of the products is less than the original rest mass of the heavy nucleus. The difference in energy is released. Setting up an experimental arrangement so that this happens is not easily done. An example of a reaction that works is the splitting of ^{235}U by a slow neutron according to the formula

$$\begin{matrix} {}^{1}_{0}n + {}^{235}_{92}U \rightarrow {}^{236}_{92}U^* \rightarrow X + Y + \text{neutrons} \end{matrix} \quad \text{(30.1)}$$

where X and Y are reaction fragment nuclei and U* is an excited nucleus, an intermediate state. There are about 90 different possibilities for reaction fragments which obey the conservation laws. The average number of neutrons released per event is about 2.5 and the average energy per event is 220 MeV.

30.2 Nuclear Reactors

The release of surplus neutrons (about 2.5 per reaction) in fission reactions makes it possible to start a chain reaction, providing these released neutrons can be made to interact with other fissionable nuclei. If ^{235}U is used as a fuel, the neutrons must be slowed down or moderated by-letting them scatter from light nuclei first. The reaction, when uncontrolled, gives us a bomb. If it is controlled, we get a nuclear reactor which can become useful as a power plant. The key steps in making a reactor are:

NUCLEAR LEAKAGE

Some of the neutrons will escape from the system and, therefore, will not contribute to producing more fissions. The smaller the system, the more neutrons, percentage-wise, will escape. It is critical to choose the right surface area to volume ratio to keep the K factor equal to 1. (K measures the number of neutrons from a reaction that will cause another reaction.)

REGULATING NEUTRON ENERGIES

Most of the neutrons from a fission event have energies so high that they are not easily captured by another ^{235}U nucleus. Therefore, they have to be slowed down, i.e., moderated. This is done by letting them scatter from light nuclei, e.g., hydrogen or carbon. After several collisions the neutrons will be going at thermal velocities. To provide for this moderation, the fuel is surrounded by graphite (carbon) or something containing a lot of hydrogen, for example, water.

NEUTRON CAPTURE

Neutrons are lost to the system if captured by nuclei other than those we want to fission. ^{238}U, for example, will absorb high energy neutrons without undergoing fission. To prevent a large loss of neutrons we need to moderate their energy as rapidly as possible.

CONTROL OF POWER LEVEL

In order to control the number of neutrons, the reactor is designed to give an excess so K > 1, and then rods of material are inserted that will absorb neutrons with a high probability. The amount of rod inserted, therefore, controls the number of available neutrons.

REACTOR SAFETY

There are several considerations in the design of nuclear reactors that must be considered to make them safe. There has been no loss of lives in the United States due to nuclear reactors. The biggest threat to health is the possible escape of radioactive materials. The rate of reactions is controlled with control rods, but there still exists the possibility of overheating because of failure in the heat extraction system or the emergency cooling system. Backups for these systems are incorporated to reduce the possibility of disaster. If one system fails for some reason, resulting in overheating, a steam explosion or a melt-down of the core of the reactor could occur, which in the worst cases could release radioactive materials in escaping steam.

30.3 Nuclear Fusion

The binding energy per nucleon increases in the light nuclei as we go up the periodic chart until we get to iron. Putting two light nuclei together to make a heavier one is called fusion and can result in the release of energy if the new fused nucleus is more tightly bound. The sun and most stars produce most of their energy through fusion. The sun fuses hydrogen to produce helium in three steps, namely

$$_1^1H + _1^1H \rightarrow _1^2H + _1^0e + \nu$$

$$_1^1H + _1^2H \rightarrow _2^3He + \gamma$$

$$_1^1H + _2^3He \rightarrow _2^4He + _1^0e + \nu \quad \text{or}$$

$$_2^3He + _2^3He \rightarrow _2^4He + _1^1H + _1^1H$$

(30.2)

FUSION REACTORS

To duplicate in the laboratory the conditions (pressure, density, temperature), found in the interior of the sun is extremely difficult. So we have looked for another way to accomplish fusion. A set of reactions that are promising are

$$_1^2H + _1^2H \rightarrow _2^3He + _1^0n \quad Q = 3.27 \text{ MeV}$$

$$_1^2H + _1^2H \rightarrow _1^3H + _1^1H \quad Q = 4.03 \text{ MeV}$$

(30.3)

$$_1^2H + _1^3H \rightarrow _2^4He + _1^0n \quad Q = 17.59 \text{ MeV}$$

where the Q values are the amounts of energy released in each reaction.

Power can be extracted from the reactor only when Lawson's criterion is satisfied, which is

$n\tau \geq 10^{14}$ s/cm^3 Deuterium-tritium interaction   (30.4)

$n\tau \geq 10^{16}$ s/cm^3 Deuterium-deuterium interaction (30.5)

n is the plasma ion density and τ is the confinement time for the plasma.

MAGNETIC FIELD CONFINEMENT

Strong magnetic fields are used to confine the plasma. The fields hold the plasma for longer sustained conditions so fusion can occur and also prevent it from touching the walls of the container, thereby losing energy and also becoming contaminated.

Our best efforts have not yet achieved break-even energy production.

30.4 Elementary Particles

It has taken a long time to understand the nature of the elementary particles. At one time the atom was considered indivisible and hence the most elemental form of matter. Then the discovery of the electron and the atomic nucleus revised that concept. The nucleus was found to be divisible into protons and neutrons and the nucleons have been found to have structure and are formed of particles we call quarks. The quarks (we know of six of them) are but one family of particles. The electron, the positron, the neutrino, etc. belong to another family called leptons. These quarks and leptons vary in mass and other important properties, but they all have the same size, zero radius, and the same intrinsic angular momentum. There are additional particles that are the primary carriers of the various fundamental forces such as the photon for the electromagnetic force and similar particles for the other forces.

30.5 The Fundamental Forces

All natural phenomena can be explained in terms of four fundamental forces, the strongest being the nuclear force. Next strongest is the electromagnetic followed by the weak, and by far the weakest is gravity. (These comparisons apply to interactions between two elementary particles upon which the forces act.)

The nuclear force is very short range (10^{-15} m) as is also the weak nuclear force, whereas both gravity and electromagetic forces extend to infinity. Gravity is about 10^{-38} times the nuclear force and the electromagnetic force is 100 times weaker than the strong force.

The nuclear force operates through the exchange of gluons, the electromagnetic force by exchanging photons, the weak interaction by the W^+, W^-, and Z^0 bosons, and gravity by the exchange of gravitons.

30.6 Positrons and Other Antiparticles

Paul Adrien Maurice Dirac in the 1920's succeeded in applying Einstein's special theory of relativity to the quantum mechanical problem of the electron. Because the total energy in relativity is given as a square root of momentum squared plus rest mass energy squared, the + and - sign in front of the square root can be interpreted as both positive and negative energy states. The negative energy states were interpreted by Dirac to represent antiparticles. Carl Anderson in 1932 discovered the antielectron (positron) and in 1955 Emilio Segrè and Owen Chamberlain discovered the antiproton, and shortly thereafter the antineutron. The antiparticle of nearly every known particle has been found since then. Some particles such as the neutral pion are their own antiparticle and spontaneously annihilate themselves. Otherwise, particles annihilate on antiparticles to give either a pair of gamma rays or different antiparticles.

30.7 Mesons and the Beginning of Particle Physics

Hideki Yukawa in 1935 predicted the existence of pi-mesons, or pions, and denoted them by the symbol π, as nuclear particles acting between nucleons to bind them together inside the nucleus. Because of the short range of the nuclear force, he was able to predict using the Heisenberg uncertainty principle as it applies to energy and time, the mass of the pion as about 200 times the electron mass. He used Δt = (range of force = 10^{-15} m)/c and $\Delta E = m_\pi c^2$. This gives us

$$m_\pi c^2 \approx \frac{hc}{2\pi d} = 2.1 \times 10^{-11} \text{ J} = 130 \text{ MeV} \qquad \text{(30.6)}$$

in good agreement with the mass of the pion.

The pions were discovered in 1947 and found to decay into another type of "electron" called the mu-meson or muon and a neutrino with a lifetime of 2.6 x 10^{-8} s. The muon then decayed into a regular electron and two neutrinos in 2.2 x 10^{-6} seconds. Thus,

$$\pi^- \rightarrow \mu^- + \bar{\nu}$$

$$\mu^- \rightarrow e^- + \nu + \bar{\nu}$$

(30.7)

where the bar above the neutrino indicates it is really an antineutrino.

The weak interaction is mediated by particles that are more massive still, so the range is very much shorter. The approximate mass of these particles is ~ 85 GeV.

30.8 Classification of Particles

HADRONS

Hadrons are particles that participate in the strong nuclear force. There are two types of hadrons, the mesons with integer spin and the baryons with half odd-integer spin. Mesons are exchanged back and forth by the baryons with the pion being the lightest meson and, therefore, the one most easily observed experimentally.

LEPTONS

There are six known leptons. They are the electron, the muon, and the tau, and a different kind of neutrino for each of these, i.e. electron neutrino, muon neutrino and tau neutrino. (Each of these has its own antiparticle). The electron neutrinos are extremely light (< 16 eV) and may have zero rest mass like the photon.

30.9 Conservation Laws

Conservation laws are extremely important in the study of elementary particles. While there are several very important conservation laws such as charge, energy, and momentum, we mention two new ones here.

BARYON NUMBER

The baryons are the hadrons with half odd-integer spin such as the nucleons, the Λ, the Σ, the Ξ, and etc. In any reaction or decay, the number of baryons on each side of the equation is constant. Antibaryons are counted with -1 for each baryon.

LEPTON NUMBER

The six leptons plus six antileptons are conserved in number in all particle interactions and decays. The lepton numbers are +1 for each lepton and -1 for each antilepton, and the total lepton number is a constant in all reactions. It has also been observed that electron-lepton number is conserved and independently also muon-lepton number and tau-lepton number.

30.10 Strange Particles and Strangeness

Hadronic processes (i.e., strong nuclear interactions) are very short range interactions with extremely short lifetimes. Yet, some known hadrons are observed with very long half-lives, relatively speaking (10^{-10}s instead of 10^{-23} s). This "strange" behavior was assumed to be due to a new quantum number like charge, but different, and was called "strangeness". Some particles were assumed to have +1 strangeness, some - 1, and some zero strangeness. Later certain particles were found to have more than one unit of strangeness, i.e. their strangeness was 2 or 3.

Strangeness is found to be conserved in number on each side of the reaction equation where the hadronic (strong nuclear) force is acting but often not in the decay. The decay is fast when strangeness is conserved, but is slow and goes by the weak force when strangeness is not conserved.

30.11 The Eightfold Way

A plot of strangeness versus charges for the baryons gives patterns involving eight of the baryons. Similar patterns are produced by such plots for the mesons but with their antiparticles included. This pattern suggested some underlying regularity which was put into a theory, called the eightfold way, involving symmetry requirements and three new particles named quarks by Murray Gell-Mann and Yuval Ne'eman in 1961. The theory predicted other patterns which were found, but sometimes with some missing particles. The "missing" particles were subsequently found, such as the Ω^-, a massive baryon with strangeness of -3.

30.12 Quarks--Finally

The eightfold way implies an underlying structure of subcomponents of hadrons, called quarks, whereas the leptons seem to be truly fundamental.

THE ORIGINAL QUARK MODEL

The first theory of quarks required only three, namely the up, the down and the strange quark. Mesons were composed of a bound quark and antiquark, and the baryons were made of three bound quarks. Every combination made a different particle. The quarks had fractional charges ($e/3$ and $2e/3$) and fractional (1/3) baryon number.

CHARM AND OTHER RECENT DEVELOPMENTS

The discovery in 1974 of a new meson called the J/ Ψ with unexpected properties led to a new quantum number something like strangeness. It was called "charm" and this led to the need of a fourth quark, one that carried charm. At this time there were only four leptons known, so with four leptons it was acceptable to also have four quarks. Then another lepton was found, the tau. Including its associated neutrino, this made six leptons in all, so two additional quarks, named the "top" and "bottom" quarks, were expected. Both of these quarks were eventually found. The announcement of the top quark discovery was made in 1994.

Quarks are never found "free." The current theories now require them always to be confined.

30.13 The Standard Model

At present we have theories that explain all elementary particle interactions and decays. These theories include the unification of the electromagnetic force with the weak force, and the theory of quarks and the hadronic interactions, called quantum chromodynamics (QCD) patterned after the quantum theory of electrodynamics (QED). These theories make up what we call the *standard model*. QCD requires recognizing that some phenomena in hadronic processes requires the addition of another type of quantum number called "color." Color is not related to visual colors, but still the colors of red, green, and blue are used for the quarks and they must be mixed in such a way as to leave every particle colorless. This requires that each quark in a baryon have a different color and for the mesons the color of the quark and an antiquark are such as to make the meson colorless.

It is the hope and belief of most particle physicists that the standard model will eventually include the unification of the electroweak force and the hadronic force into a "grand unified theory" (GUT).

30.14 The Cosmic Connection

At very high energies we observe the particle reactions discussed above. These conditions generally do not exist in the everyday world, but rather in gigantic particle accelerators. However, there is evidence for an exception to this situation, namely the universe is expanding and cooling. Extrapolating backwards gives us ever increasing compaction and increasing temperature of the universe. Extrapolating backwards far enough gives us conditions we find in the particle accelerators and beyond. Thus particle physics has a connection to the earliest moments in the evolution of the universe.

OBSERVATRION OF RADIATION FROM THE PRIMORDIAL FIREBALL

In 1965 Arno A. Penzias and Robert W. Wilson of Bell labs observed a universal radiation coming from the universe that is recognized today as the radiation from the primordial fireball, i.e. the Big Bang of the universe when it all began. Originally the temperature was exceedingly high so there was ample energy to make every conceivable particle reaction possible and a great deal of very high energy Black Body radiation was emitted. Due to the expansion and subsequent cooling, that radiation corresponds to a Black Body of 2.9 K temperature today.

30.15 Concept Statements and Questions

1. Nuclei are extremely small compared to the atoms and are made of protons and neutrons.

2. The density of all nuclei is the same and hence the radius is proportional to the atomic number, A.

3. Nuclei have intrinsic angular momentum called nuclear spin. Both integer and half-odd integer values for spin are possible. Zero is very common.

4. The binding energy of nuclei is so great that there is a measurable difference in mass between the nucleus and the sum of its constituents when not bound.

5. The decay rate of radioactive material is proportional to the number of radioactive atoms present.

6. The particles emitted in radioactive decay are alpha particles, beta particles, and gamma rays, unless the nucleus is highly excited, whereupon the emission of other particles is possible.

7. The Q of a reaction tells how much energy is involved in the reaction.

8. There are very many subatomic particles in nature and each has an antiparticle. Some particles are their own antiparticle.

9. There are four fundamental forces in nature presently known.

10. The particles that interact through the strong nuclear force are called hadrons. There are two types of hadrons. What are they?

11. The mass of pions is estimated through the uncertainty principle for energy and time.

12. The family of leptons has 6 members. What are they? How many quarks are there (excluding color differentiation)?

13. What conservation laws pertain to baryons? to leptons? to strange particles? to particles with charm?

14. How does cosmic blackbody radiation provide evidence that the universe is expanding?

30.16 Hints for Solving the Problems

General Hints

1. In particle reactions or decays we can strictly apply the conservation of energy and momentum and also baryon number, charge, and lepton number.

2. To check on conservation of strangeness and apply it, the reaction considered must be hadronic and many decays are not. This is also true for charm, topness and bottomness.

3. Check the tables for the properties of each particle.

4. Mass and energy are equivalent so in particle-antiparticle annihilation or production, energy such as contained in a photon (no rest mass) can be totally changed to a particle with rest mass and vice versa. Since momentum has to be conserved, the annihilation of a particle and antiparticle has to yield two photons or occur near a third object which can take up the momentum.

Hints for Solving Selected Problems

1-11. Balance the nuclear reactions as needed. Figure the Q of the reaction from the masses involved. Apply conservation of energy and convert energy to frequency through $E = hf$. From the energy of the reaction convert to energy output or power output and use.

12-19. Apply the same hints as Problems 1-11 but now you will be considering fusion events instead of fission events.

20-21. When particle-antiparticle systems annihilate into two photons, each photon carries off the rest mass energy of one of the particles. Apply conservation of energy to use energy-mass equivalence.

22-25. Use lepton number, muon number and electron number conservation in Problem 23. In Problem 25, assume conservation of energy. For 22 use the uncertainty principle for energy and time.

26-39. Apply the conservation laws from the tables in the text. Look for conservation of baryon number, lepton number, charge, strangeness, mass (energy), electron-lepton number, muon-lepton number, etc. A hadron can decay via the weak nuclear interaction and not conserve strangeness.

41-46. Apply known properties of the quarks combined with the conservation laws.

For Further Study in Chapter 7

More on Gravity

Gravitational Potential Energy

When the change in elevation is small, we can compute the gravitational potential energy from the formula $PE = mgh$. This is only an approximation formula because we are assuming that g is constant. A more exact expression which accounts for changes in g as the distance, r, from the center of the earth changes is given by

$$PE = -G\frac{M_E m}{r} \qquad \text{(7a. 1)}$$

Escape Speed

The total mechanical energy of an object of mass m is the sum of the kinetic energy and the potential energy, that is

$$KE + PE = \frac{1}{2}mv^2 - G\frac{M_E m}{r} \tag{7a. 2}$$

If the total energy is zero at infinity, i.e., the object is at rest where the potential energy is zero, then the kinetic energy is equal to the gravitational potential energy and the object has enough energy to escape. The speed necessary for that to happen is called the escape speed and can be calculated by solving for v in the above equation when $r = R_E$, so

$$v_{esc} = \sqrt{\frac{2GM_E}{R_E}} \tag{7a. 3}$$

The escape speed from the earth's surface is about 11.2 km/sec which equals about 25,000 mi/h.

Gravity in the Extreme, Black Holes

The birth of stars occurs in large gaseous clouds called nebulae. Gravity pulls the gas together in various sized clumps. As it does so the clumps get smaller in diameter and their temperatures increase because of release of gravitational energy. The protostars become true stars when the central temperatures reach several million degrees and nuclear fusion begins in the core. The star is said to be born. After all the "fuel" for fusion processes is consumed, the core collapses and a runaway explosion occurs which blasts off the outer layers of the star. This explosion is called a supernova. The core continues to collapse and if massive enough it becomes so small and condensed that the escape speed equals or exceeds the speed of light. It becomes a black hole.

Newtonian gravity and laws of motion are inadequate to properly treat this extreme gravitational situation, but, even so, the equation for the maximum or critical radius (called the Schwartschild radius) for which the black hole can exist for a given mass, M, is correctly computed from our escape speed formula when $v_{esc} = c$, so that

$$R = \frac{2GM}{c^2}$$ (7a. 4)

R for the Earth is about the size of a tennis ball and R for the sun is about 3.0 km.

Hints for Solving the Problems

1. Set the centripetal force equal to the gravitational force at the radius r to get the speed from which the period and kinetic energy can be calculated.

3. Use the potential energy formula for (a). Once you know the potential energy, by putting in an extra r in the demoninator, you have the force.

4. Substitute values from the table into the escape speed formula.

5. Get g from $g = GM/r^2$. Weight is mg.

For Further Study in Chapter 9

Solids and Fluids

Surface Tension

Small droplets of a liquid are usually spherical. This is because the cohesion of the surface molecules results in a force per unit length, F/L, or an energy per unit surface area called surface tension. The energy of the system is minimized when the surface is spherical. The tension, γ, is thus

$$\gamma \equiv \frac{F}{L} \tag{9a. 1}$$

To measure surface tension a wire of length L, usually a hoop, is pulled vertically through the surface and the surface tension is

$$\gamma = \frac{F}{2L} \tag{9a. 2}$$

where the factor of two is included since the liquid acts on both sides of the wire, making the total length $2L$.

γ decreases with temperature and can be greatly reduced by adding certain ingredients such as detergents to the liquid.

The Surface of Liquids

Forces between like molecules are called *cohesive* forces and between unlike molecules they are called *adhesive*. In terms of these forces we can explain why a liquid like water goes up on a piece of glass whereas mercury goes down. In water the adhesive forces of the water for the glass are greater than the cohesive forces of water for water. For mercury, on the other hand, the cohesive forces are greater.

Capillary Action

The rise of a liquid in a capillary tube (a tube with a very small diameter opening) is understandable in terms of surface tension and the difference between the adhesive and cohesive forces. The liquid will rise in the tube to a height where the vertical component of the surface tension force, $F_v = \gamma(2\pi r)cos\phi$, equals the weight of the fluid column above the surface, $w = Mg = \rho V g = \rho g \pi r^2 h$. The height, h, therefore is

$$h = \frac{2\gamma}{\rho g h}\cos\phi \qquad \text{(9a. 3)}$$

A capillary tube inserted into a fluid like mercury will push the liquid down by an amount calculated similarly to the above equation.

Viscosity

Viscosity refers to the internal friction of a fluid as one layer moves relative to another. Thus viscosity can be related to sheer stresses, F/A, and shear strains, $\Delta x/L$. The coefficient of viscosity, η, is defined as the ratio of the shearing stress to the time rate of change of the shear strain. Thus for η we get

$$\eta \equiv \frac{FL}{Av} \tag{9a. 4}$$

The SI unit of viscosity is $N \cdot s/m^2 = 1$ poise.

Poiseuille's Law

Poiseuille's law gives the rate of flow of a fluid through a tube of length, L, when there is a pressure difference, $P_1 - P_2$, and the viscosity is η. The law, applied to incompressible fluids, is

$$\text{Rate of flow} = \frac{\Delta V}{\Delta t} = \frac{(P_1 - P_2)\,(\pi R^4)}{8L\eta} \tag{9a. 5}$$

Reynolds Number

When the fluid flow changes from simple streamline flow to turbulent flow, the motion is very different. The onset of turbulent motion is determined by the dimensionless factor called the Reynolds number, given by

$$RN = \frac{\rho vd}{\eta} \tag{9a. 6}$$

where ρ is the density, v is the average speed, d is the diameter of the tube, and η the viscosity. If RN is below 2000, the flow is streamline, if it is about 3000, the flow is turbulent, and between 2000 and 3000 the flow is unstable.

Transport Phenomena

In addition to pressure differences causing a fluid to flow, concentration differences will also. Two fundamental processes that account for this flow are *diffusion* and *osmosis*.

DIFFUSION

Diffusion of a fluid occurs due to random mixing which tends to achieve maximal disorder. The diffusion rate is known as Fick's law and is given by

$$\text{Diffusion rate} = \frac{\text{mass}}{\text{time}} = \frac{\Delta M}{\Delta t} = DA\left(\frac{C_2 - C_1}{L}\right) \quad \text{(9a. 7)}$$

C is concentration, A is area, and D is a constant of proportionality.

THE SIZE OF CELLS AND OSMOSIS

Osmosis is defined as the movement of water from a region where its concentration is high, across a selectively permeable membrane, into a region where its concentration is lower. The process continues until the concentrations on the two sides of the membrane are equal.

Cells are of such a small size that their surface areas are large compared to their volume. It is across the cell membrane that osmosis takes place. Through osmosis plants have the ability to move water from the roots to the tops of the trees.

Motion through a Viscous Medium

The resistive force an object experiences in moving through a fluid is basically viscous drag. The amount of force depends upon the size, the speed and the viscosity and follows Stokes' law, which is

$$F_r = 6\pi\eta rv \qquad \text{(9a. 8)}$$

Besides the viscous force we have the weight and the buoyant force acting on the object. Putting this all together for vertical fall we get, for a spherical object of radius r and density ρ, falling through a fluid of viscosity η and density ρ_f,

$$6\pi\eta rv_t + \rho_f g \left(\frac{4}{3}\pi r^3\right) = \rho g \left(\frac{4}{3}\pi r^3\right) \qquad \text{(9a. 9)}$$

which when solved for the terminal speed v_t, gives

$$v_t = \frac{2r^2 g}{9\eta}(\rho - \rho_f) \qquad \text{(9a. 10)}$$

SEDIMENTATION AND CENTRIFUGATION

If the object is not spherical we still get weight = buoyant force + viscous force, but we can no longer use Stokes' law for the viscous force, but rather $F_r = kv$. The terminal speed is now expressed as

$$v_t = \frac{mg}{k}\left(1 - \frac{\rho_f}{\rho}\right)$$ (9a. 11)

This terminal velocity is sometimes called the *sedimentation rate*. To increase the rate for a given object and fluid we can increase the effective g by placing everything in a centrifuge. Now g becomes $\omega^2 r$.

Hints for Solving the Problems

General Hints

1. Study the examples in the text carefully.

Hints for Solving Selected Problems

1-4. Use the surface tension formula. Remember to include the factor of 2 on the length when pulling a wire through a liquid.

5-9. Study Example 9a.2 in the text and use Equation 9a.3 in this guide.

10-23. Look at Examples 9a.3 amd 9a.4 in the text.

For Further Study in Chapter 15

Electric Forces and Electric Fields

We introduce in this chapter an elegant way for determining the electric field when we have symmetric distributions of charge. The method was developed by Gauss and is known as Gauss' law and builds around the concept of electric flux.

Electric Flux

Electric flux is defined as the electric field penetrating a surface area A so the field is perpendicular to the plane of the surface. We have then

$$\Phi = EA \qquad \text{(15a. 1)}$$

If the perpendicular to the plane of the surface makes an angle with respect to the electric field, then the flux becomes

$$\Phi = EA\cos\theta \qquad \text{(15a. 2)}$$

If the flux lines pass through an enclosed surface, we adopt the convention that those lines going into the surface are negative and those going out are positive.

Gauss' Law

The electric flux from a particle with charge Q is proportional to Q. That is Gauss' law. For a vacuum or air we can identify the proportionality constant as $\varepsilon_0 = 1/4\pi k$. We can put this together and for a closed surface we get

$$\Phi = \sum EA\cos\theta = \frac{Q}{\varepsilon_0} \qquad \text{(15a. 3)}$$

Using Gauss' law we can readily get the magnitude of the electric field surrounding a point charge q. We do this by picking a spherical gaussian surface symmetrically centered on the charge. Thus **E** on the surface is everywhere perpendicular to the surface, so $EA\cos\theta$ can be written as $E(4\pi r^2)$. Since $Q = q$ we can write

$$E4\pi r^2 = \frac{q}{\varepsilon_0} \quad \text{or} \quad E = \frac{q}{4\pi\varepsilon_0 r^2} = k\frac{q}{r^2} \qquad \text{(15a. 4)}$$

Examples found in the text show other ways in which Gauss' law can be applied to simplify certain problems.

Hints for Solving the Problems

1. Multiply E times $A\cos\theta$ for part (a).and (b).

2, Find E from Φ/A.

3. Flux equals q/ε_0 for a surface surrounding a charge q..

4. A gaussian sphere drawn outside the sphere of charge encompasses how much charge? If drawn inside how much does it encompass?

5. Note that $q/A = \sigma$.

6. Draw the gaussian surface just below the charged surface.

APPENDIX A

Table A.1
SI BASE UNITS

Base Quantity	SI Base Unit Name	Symbol
Length	Meter	m
Mass	Kilogram	kg
Time	Second	s
Electric current	Ampere	A
Temperature	Kelvin	K
Amount of substance	Mole	mol
Luminous intensity	Candela	cd

Table A.2
Mathematical Symbols Used in the Text and Their Meaning

Symbol	Meaning
$=$	is equal to
$\neq$	is not equal to
$\equiv$	is defined as
$\propto$	is proportional to
$>$	is greater than
$<$	is less than
$\gg$	is much greater than
$\ll$	is much less than
$\sim$	is approximately equal to
Δx	change in x
Σx_i	sum of all quantities x_i
$|x|$	magnitude of x (always a positive quantity)

Table A.4
The Greek Alphabet

Alpha	A	α	Iota	I	ι	Rho	P	ρ
Beta	B	β	Kappa	K	κ	Sigma	Σ	σ
Gamma	Γ	γ	Lambda	Λ	λ	Tau	T	τ
Delta	Δ	δ	Mu	M	μ	Upsilon	Y	ν
Epsilon	E	ϵ	Nu	N	ν	Phi	Φ	ϕ
Zeta	Z	ζ	Xi	Ξ	ξ	Chi	X	χ
Eta	H	η	Omicron	O	o	Psi	Ψ	ψ
Theta	Θ	θ	Pi	Π	π	Omega	Ω	ω

Table A.3
Standard Abbreviations of Units

Abbreviation	Unit	Abbreviation	Unit
A	ampere	J	joule
Å	angstrom	K	kelvin
atm	atmosphere	kcal	kilocalorie
Btu	British thermal unit	kg	kilogram
C	coulomb	km	kilometer
°C	degree Celsius	kmol	kilomole
cal	calorie	lb	pound
cm	centimeter	m	meter
deg	degree (angle)	min	minute
eV	electron volt	N	newton
°F	degree Fahrenheit	rev	revolution
ft	foot	s	second
G	gauss	T	tesla
g	gram	u	atomic mass unit
H	henry	V	volt
h	hour	W	watt
hp	horsepower	Wb	weber
Hz	hertz	μm	micrometer
in.	inch	Ω	ohm

Table A.5
Derived SE Units

Quantity	Name	Symbol	Expression in Terms of Base Units	Expression in Terms of Other SI Units
Plane angle	Radian	rad	m/m	
Frequency	Hertz	Hz	s^{-1}	
Force	Newton	N	$kg \cdot m/s^2$	J/m
Pressure	Pascal	Pa	$kg/m \cdot s^2$	N/m^2
Energy: work	Joule	J	$kg \cdot m^2/s^2$	$N \cdot m$
Power	Watt	W	$kg \cdot m^2/s^3$	J/s
Electric charge	Coulomb	C	$A \cdot s$	
Electric potential (emf)	Volt	V	$kg \cdot m^2/A \cdot s^3$	W/A
Capacitance	Farad	F	$A^2 \cdot s^4/kg \cdot m^2$	C/A
Electric resistance	Ohm	Ω	$kg \cdot m^2/A^2 \cdot s^3$	V/A
Magnetic flux	Weber	Wb	$kg \cdot m^2/A \cdot s^2$	$V \cdot s$
Magnetic field intensity	Tesla	T	$kg/A \cdot s^2$	Wb/m^2
Inductance	Henry	H	$kg \cdot m^2/A^2 \cdot s^3$	Wb/A

INDEX